Dedicated to My Parents

fundamental concepts of modern mathematics

max d. larsen, university of nebraska

addison-wesley publishing company
reading, massachusetts
menlo park, california · london · don mills, ontario

 Printed in the United States of America. Published simultaneously in Canada. Library of Congress Catalog Card No. 77-87037.

preface

This book is designed as a text for an introductory course in symbolic logic and the theory of sets. It grew out of the author's realization that many beginning college mathematics students are completely unfamiliar with the notions of symbolic logic and elementary set theory. The material in this text is taught at the University of Nebraska in a course which attempts to bridge the gap between the cookbook mathematics of calculus and the beginning courses in abstract algebra and modern geometry in which the student is expected to understand and to write proofs.

Chapter 1 is an introduction to symbolic logic. The usual connectives are introduced first on an intuitive level via truth tables. The connectives are then defined axiomatically. This forces the idea of proof to come up early. The extent to which the student masters the idea of proof depends largely upon his exposure to it; hence it seems important to introduce the student to the idea of proof without the confusion raised by other concepts.

In Chapter 2 the theory of sets is introduced, and the major concepts of set theory needed for a course in modern algebra or geometry are discussed. The topics covered are mappings, cardinality, equivalence relations, binary operations, and order relations. Throughout Chapters 1 and 2 there are many exercises to strengthen the student's understanding of the concepts involved. At the end of the book there is a short section of answers and hints for selected exercises.

The concepts from symbolic logic and the theory of sets remain sterile and isolated unless they are used. Consequently, three possible choices of subject matter have been included in this book for the purpose of applying the knowledge gained in Chapters 1 and 2. These choices are offered in Chapters 3, 4, and 5. These chapters are independent of each other. Usually any one of these chapters along with the first two chapters is sufficient material for a one-semester course.

The number systems through the complex numbers are developed from Peano's Postulates in Chapter 3. The construction of the number systems is an item often omitted in many curricula, but every mathematics student—especially those who plan to teach—should be exposed to the development of the number systems. The first three chapters of this book contain the material for a course which will provide this exposure while presenting the basic concepts of modern mathematics. The proofs of many of the theorems of Chapter 3 are left as exercises which comprise a vital part of this chapter, since they require that the student use the notions of symbolic logic and set theory in the construction of his proofs.

Chapter 4 consists of a rudimentary treatment of properties of the integers. With the assumption that the positive integers are well-ordered, this chapter explores divisibility properties of the integers and ends with the proof of the unique factorization theorem. The exercises in this chapter are especially easy to handle, since the students have a certain familiarity with the concepts involved; but, on the other hand, the exercises do stress the idea of proof and many are of the form "prove or disprove." If a course in modern algebra is to follow the course for which this text is used, the results of this chapter will prove helpful.

As an example of an abstract mathematical system, a chapter on groups is included. Chapter 5 is nearly the same as the chapter on groups in the author's *Introduction to Modern Algebraic Concepts* (Addison-Wesley, 1969). This chapter covers group theory through Lagrange's Theorem and factor groups and has a section on homomorphisms. If a course in modern algebra is to follow, normally one would not do Chapter 5. But, if this book is used as a text for a course for nonmathematics majors who will not be taking courses in abstract algebra, Chapters 1, 2, and 5 provide an introduction to modern abstract mathematics.

There are no formal prerequisites for understanding the material of this book. However, a sincere interest in mathematics and a desire to learn are essential. The concepts presented in the book are important to the nonmathematician as well as to the mathematician. The diversity of the three independent chapters following the symbolic logic and set theory place the limits on the use of this text in the hands of the teacher.

I wish to express my sincere thanks to Richard J. Shumway for reading the manuscript and for making many helpful suggestions.

Lincoln, Nebraska
August, 1969

M. D. L.

note to the student

For many of you, this course will be your first encounter with modern mathematics. Consequently, a word concerning your approach to the concepts presented in this book may be worthwhile. If this is your first dealing with mathematics from a modern point of view, you will no doubt find it quite different from the other courses you have had in mathematics, such as calculus and differential equations. Many students of mathematics as well as many professional mathematicians find the purity of the modern abstract approach quite refreshing. I certainly do, and I sincerely hope you can experience some of the excitement and joy that I find in mathematics.

The only requirement made of you is that you have an alert and creative mind and that you *work*. Except in extreme cases, the material presented here will remain dead for you unless you are able, through your own work and participation in the course, to bring life to these exciting concepts. You are past the stage in your mathematical training in which it is sufficient to memorize a formula and apply it to various problems. At this stage you must *think*, and for many of us that means *work*.

A majority of the activities asked of a student studying abstract algebra consists of writing proofs of the various theorems. For many of you, this will be a new experience. In general there is no prescribed or established pattern to follow in writing proofs. Mark my words, you will soon be complaining, "I do not know how to get started writing the proof." There is no answer to your dilemma except to try every way you can think of until you find one that works. By a careful consideration of the facts that you know, the hypothesis of the theorem you are trying to prove, and the conclusion for which you are striving, you should be able to ferret out the essential points and thus have a basis from which to "get started." If you try one approach and it does not seem to work, abandon it and try another. Do not give up simply because the first try fails. As you gain experience, your first approach will fail less often. This "experience" to which I refer comes from working on proofs until you succeed in writing a correct proof even though you may suffer several false starts.

Your next question will probably be "How do I know my proof is correct?" For this question, there is an easy answer. The fact that this question has an easy answer is one major reason for the appeal of mathematics. Every system of mathematics is based on certain logical principles, on undefined terms and definitions based on them, and on certain axioms. If, at every step of your proof, you comply with the rules of logic and with the definitions and axioms of your system, then your proof will be correct. Hence the validity of a proof depends solely upon the validity of each individual step.

With a concentration on proving each theorem, there is a danger of "missing the forest because of all the trees." One aspect of the treatment of modern mathematics in this book that must not be missed is the method one employs in investigating a certain algebraic system. There are certain questions that one usually asks concerning the properties of a mathematical system which aid in the development of the theory associated with the system. You must be aware of these questions. At the conclusion of your work with the systems in this book, you should be able to begin with the definitions and axioms of a completely new system and develop the properties of it.

Let me close this note with an anecdote which sums up what I have been saying. At the beginning of each semester of classes at The University of Kansas, Professor G. Baley Price, the chairman of the mathematics department, has a meeting of all staff members. At this meeting, he reminds all that are present—graduate teaching assistants and professors alike—that "Mathematics is not a spectator sport."* The wisdom in these words becomes more evident to me the longer I work in mathematics. If you have an honest interest in mathematics, then get out of the grandstand and get to *work*! The world of mathematics is yours if you are willing to be one of the "players."

M. D. L.

* Professor Price gives John L. Kelley, Professor of Mathematics at the University of California, credit for originating this statement.

table of contents

1 elementary logic

1.1 INTRODUCTION

Logic is the theory of systematic reasoning.* Symbolic logic then is the formal theory of logic. The logic which we will consider here is rather elementary and is the logic which is commonly used in mathematics and other sciences.

The study of logic is important in its own right, since systematic reasoning is the basis of all one does in mathematics. However, here we are interested in logic mainly as it will be used later in the study of mathematics. First, the symbols and laws we develop in this chapter will be used to simplify and clarify later work. Also, the axiomatic approach used in the study of logic here illustrates the nature of a formal mathematical theory. The student should become aware of the wide applicability of the concepts and notation of symbolic logic as well as the logical laws which will be proved.

1.2 AN INTUITIVE INTRODUCTION

The purpose of this section is to introduce in an intuitive manner the logical formulas which will be used later in the axiomatic development. Truth tables are used to illustrate the meaning of the several connectives. Since we are striving for applicability of symbolic logic, it is important that the student understand the meaning of the connectives well enough to translate verbal statements into logical formulas and logical formulas into acceptable verbal statements.

A *proposition* is a declarative sentence which is either true or false, *but not both.* For example, "Mary went to town" is a proposition; but

---

* Kenneth O. May, *Elements of Modern Mathematics* (Addison-Wesley, 1959).

"Study mathematics whenever you can" is not. Also "$2 + 6 = 10$" is a proposition which is in fact false; but since the truth or falsity of "$x + 6 = 10$" is unknown unless x is specified, "$x + 6 = 10$" is not a proposition. Lower-case Latin letters, such as $p, q, r, \ldots$, are usually used as symbols for propositions.

Compound sentences are composed of one or more simple sentences and certain connectives joining the simple sentences. In the same way we can join simple propositions by connectives and obtain a new proposition. If p and q are propositions, we shall be interested in the following propositions formed from p and q and connectives which we will define later: "not-p," "p and q," "p or q," "p or else q," "if p, then q," and "p if and only if q."

A *truth table* of a logical formula shows the conditions under which the logical formula is true and those under which it is false. Truth tables are not an integral part of symbolic logic, and they are used here merely as a convenient device to present the reader with an intuitive feeling for the various connectives. Truth tables can be used, however, to determine the conditions under which a certain logical formula is true or false. If one sets up truth tables as a criterion for establishing the truth or falsity of certain logical formulas, one can use them as a device for proofs. However, we have chosen an axiomatic approach to symbolic logic and will use truth tables only as an informal tool for understanding.

The formal name for the connective *not* is *negation*, and the symbol for it is $\sim$. If p is a proposition, then $\sim p$ (read "not-p") will be its negation. The proposition $\sim p$ is false whenever p is true, and p is false whenever $\sim p$ is true. These conditions are shown in the following truth table:

p	$\sim p$
T	F
F	T

Note that $\sim$ is meaningful when used with only one proposition. This is not true of the other connectives.

The symbol for *and* is $\wedge$, and its formal name is *conjunction*. Thus if p and q are propositions, $p \wedge q$ (read "p and q") is the conjunction of p and q. Naturally the truth of $p \wedge q$ depends on the truth of p and the truth of q. We say that $p \wedge q$ is a true proposition only when both p and q are true; if either p or q is false, then $p \wedge q$ is false. This leads us to the following truth table, which serves as a definition of $\wedge$:

p	q	$p \wedge q$
T	T	T
T	F	F
F	T	F
F	F	F

In the English language, the connective *or* has more than one meaning. For instance, the exact meaning of the sentence "Mary or Ann went to town" is indefinite. This sentence can mean two different things, depending upon the context in which it is used. It could mean that one of the girls went to town and maybe both did, or it could mean one went, but not both. In symbolic logic we must remove this ambiguity. The connective *or* is denoted by $\vee$ and, if p and q are propositions, we say that $p \vee q$ is the *disjunction* of p and q. The proposition $p \vee q$ is true whenever p is true or q is true (and thus $p \vee q$ is true whenever both p and q are true) and false when both p and q are false. The truth table which illustrates the meaning of $\vee$ is

p	q	$p \vee q$
T	T	T
T	F	T
F	T	T
F	F	F

When one means that p or q is true, but not both, then the *exclusive disjunction*, denoted by $\veebar$, is used. The sentence $p \veebar q$ is read "p or else q" and has the following truth table:

p	q	$p \veebar q$
T	T	F
T	F	T
F	T	T
F	F	F

The next connective we will consider is the *conditional* or *implication*, which is denoted by $\Rightarrow$. If p and q are propositions, then $p \Rightarrow q$ is read "if p, then q" or "p implies q." Let us consider an example. Let p stand for "It is raining" and let q stand for "I will stay home." Then $p \Rightarrow q$ is read "If it is raining, then I will stay home." What must I do to violate this sentence? Clearly it is violated if I go out when it is raining, or in other words, if it rains and I do not stay home. Now suppose it does not rain. Am I free to do anything I like and still not violate the sentence? Yes! Consequently, $p \Rightarrow q$ is false only when p is true and q is false. The conditional $p \Rightarrow q$ makes no claims on what must follow when p is false; it is only when p is true that we make claims on q. In terms of a truth table then, the truth value of $p \Rightarrow q$ is given by

p	q	$p \Rightarrow q$
T	T	T
T	F	F
F	T	T
F	F	T

Other English equivalents for $p \Rightarrow q$ are "q whenever p," "p is a sufficient condition for q," "q is a necessary condition for p," and "p only if q." By considering specific examples, the reader should convince himself that the above English expressions have the same meaning as the one we have assigned to $p \Rightarrow q$.

The last connective we will consider is the *biconditional* or *equivalence*, denoted by $\Leftrightarrow$. If p and q are propositions, $p \Leftrightarrow q$ is read "p if and only if q." The expression $p \Leftrightarrow q$ is true when both p and q are true and when both p and q are false, and is false otherwise. The truth table for $p \Leftrightarrow q$ is

p	q	$p \Leftrightarrow q$
T	T	T
T	F	F
F	T	F
F	F	T

Using the six connectives $\sim$, $\wedge$, $\vee$, $\veebar$, $\Rightarrow$, and $\Leftrightarrow$, we can produce a multitude of propositions by combining simple propositions by means of one or more of these connectives. For instance, if p, q, and r are propositions, we might consider the proposition $(p \wedge q) \Rightarrow r$ and construct its truth table as follows:

p	q	r	$p \wedge q$	$(p \wedge q) \Rightarrow r$
T	T	T	T	T
T	T	F	T	F
T	F	T	F	T
T	F	F	F	T
F	T	T	F	T
F	T	F	F	T
F	F	T	F	T
F	F	F	F	T

A *theorem* (or *lemma*) is a proposition which is always true. Sometimes a logical formula or symbolic sentence which is a theorem is called a *tautology*. The negation of a theorem or tautology is called a *contradiction*; i.e., a contradiction is a proposition which is always false. One informal way to check to see whether or not a certain logical formula is a theorem is to construct its truth table. However, for an actual proof we will need the formal definitions and axioms to be stated later.

1.2.1 Exercises

1. Using the connectives discussed in this section, translate the following statements into symbolic language. Let p, q, and r denote propositions.

   a) p, when q but not r.
   b) p when q, but not r.
   c) r never happens when q, but always happens when p.
   d) p is a sufficient condition for q.
   e) p, q, and r cannot be true simultaneously.
   f) p only if q.
   g) Either p or r will produce q.
   h) p is a necessary condition for q.
   i) q if p.
   j) p unless q.

2. Express the following sentences in symbolic language.

   a) It is not necessary to be rich to be happy.
   b) Joe will win if he is better than Tom.
   c) There is no fire without smoke.
   d) In a good marriage it is not obvious who the boss is.
   e) Mary will be here provided Joe gave her the message.
   f) Red dots are a symptom of measles.
   g) In order to be successful, one must work hard.
   h) I mean what I say.
   i) I say what I mean.
   j) He must be stupid not to understand that.

3. Let p, q, r, and s be propositions. Form truth tables for the following.

   a) $p \Rightarrow (q \vee r)$. b) $(\sim p \vee q) \Rightarrow \sim q$. c) $(q \vee p) \Leftrightarrow \sim p$.
   d) $(p \wedge s) \vee r \Rightarrow q$. e) $(r \Rightarrow p) \vee (p \Rightarrow \sim q)$.

4. The connectives $\sim$, $\wedge$, $\vee$, $\veebar$, $\Rightarrow$, and $\Leftrightarrow$ are not independent; that is, there are certain relations between them. For instance, $p \vee q = \sim(\sim p \wedge \sim q)$.

   a) Express all the connectives in terms of $\wedge$ and $\sim$.
   b) Express all the connectives in terms of $\Rightarrow$ and $\sim$.

1.3 FORMAL DEFINITIONS

In the development of any mathematical or logical system, one begins with *undefined terms*. Many common approaches to plane geometry take *point*, *line*, and *plane* as undefined terms. The undefined terms in a system are used as a basis upon which are built the terminology and language of the system. Each new term must be defined by using either the undefined terms or terms which have already been defined with them. In fact, in order to do things correctly, there really should be a list of undefined terms at the beginning of every dictionary.

There are several possible choices for undefined terms here. We choose $\vee$, $\sim$, and $=$ to be undefined. Letting lower-case Latin letters denote propositions, we make the following definitions.

1.3.1 Definition

a) $p \wedge q = \sim(\sim p \vee \sim q)$.
b) $p \Rightarrow q = \sim p \vee q$.
c) $p \Leftrightarrow q = (p \Rightarrow q) \wedge (q \Rightarrow p)$.
d) $p \veebar q = (p \vee q) \wedge \sim(p \wedge q)$.

We will allow as propositions p, $\sim p$, $p \vee q$, $p \wedge q$, $p \Rightarrow q$, $p \Leftrightarrow q$, $p \veebar q$, and any expression obtained by replacing the p's and q's by any other proposition.

Next we must list some axioms which will govern our manipulations of these definitions. The axioms are natural and not at all unreasonable.

1.3.2 Axioms

a) $p = p$ (identity)
b) $p \vee q = q \vee p$ (commutativity of $\vee$)
c) $p \vee (q \vee r) = (p \vee q) \vee r$ (associativity of $\vee$)
d) $p \vee (q \wedge r) = (p \vee q) \wedge (p \vee r)$ (distributivity of $\vee$ over $\wedge$)
e) $p \vee p = p$
f) $\sim\sim p = p$

Next we must give some *rules of proof* which state what procedures are allowable in constructing proofs. They are very simple and are stated formally here for completeness. In any formal mathematical system, one selects certain undefined terms and then gives axioms and definitions from which one must obtain theorems. But how shall one derive these theorems? By stating rules of proof, we are stating in advance what will be acceptable methods for deriving theorems from the axioms and definitions. Then we can claim that our results are derived from the axioms and definitions according to these rules. Consequently, anyone who finds our axioms and rules acceptable will believe our results. Moreover, in any situation in which the axioms are true, the theorems can be applied to that situation.

For ease in expressing these rules of proof, let us say that two expressions a and b are *equal* if $a = b$.

1.3.3 Rules of Proof

a) ***Rule of Replacement.*** *Any term in a logical formula may be replaced by an equal term.*

b) ***Rule of Substitution.*** *A sentence which is obtained by substituting propositions for the terms of a theorem is itself a theorem.*

c) ***Rule of Inference.*** *If a conditional statement $p \Rightarrow q$ is a theorem and if the hypothesis p is a theorem, then the conclusion q is a theorem.*

The reader must read these rules carefully and take care to distinguish between them. Reference to the rule used will be made in formal proofs. *Replacement* is merely the process of putting one of two equal terms in place of another. Note that replacement requires that the terms interchanged be equal. To indicate that a term x is replaced by a term y, one writes $(x : y)$. Thus, if the replacement of x by y is the reason for the validity of a step in a formal proof, one shows this by writing "*Rep* $(x : y)$"; for instance, if $q = r$,

$$p \wedge q = p \wedge r \qquad Rep\,(q : r).$$

Substitution involves the interchange of propositions or statements in a theorem. Substitution does not necessarily involve equality. The notation for indicating substitution is similar to that for replacement and is illustrated below:

$$(p \Rightarrow q) \vee q = q \vee (p \Rightarrow q) \qquad Sub\,(p : p \Rightarrow q), \quad 1.3.2(b).$$

Here we have substituted $p \Rightarrow q$ for p in 1.3.2(b). When substituting an expression for another in a theorem, the new expression must be inserted everywhere the old expression appears. On the other hand, a symbol may *replace* an equal symbol in just one of its appearances in an expression.

The *rule of inference* is one which we will use to "apply" the theorems we have already proved. Suppose that for some reason, possibly a previous result, we know that p is a theorem. If we can prove that $p \Rightarrow q$ is a theorem, then we can conclude that q is a theorem. Formally this proof would be written as follows:

(1)	$p \Rightarrow q$	
(2)	p	(From previous results)
	q	*Inf* (1), (2).

The notation here means that q is a theorem by the rule of inference applied to theorems (1) and (2).

1.4 LOGICAL IDENTITIES

Now that we have established the foundations for our logical system, we are ready to begin to prove the theorems which make up our theory. There are many orders in which the theorems may be proved. The essential thing is that every theorem be dependent only upon theorems we have previously established. Since our axioms are stated primarily in terms of $\vee$, one natural approach would be to try to prove theorems about $\wedge$ which are similar in form to the axioms.

Two theorems which are named after the English mathematician Augustus De Morgan (1806–1871) are useful in establishing relations between $\vee$ and $\wedge$.

1.4.1 $\sim(p \vee q) = \sim p \wedge \sim q.$

(De Morgan's laws)

1.4.2 $\sim(p \wedge q) = \sim p \vee \sim q.$

To prove 1.4.1, note that 1.3.1(a) with *Sub* $(p : \sim p,\ q : \sim q)$ is $\sim p \wedge \sim q = \sim(\sim\sim p \vee \sim\sim q)$. However, by 1.3.2(f), $\sim\sim p = p$ and $\sim\sim q = q$. Hence, $\sim p \wedge \sim q = \sim(p \vee q)$.

The style of the above proof is quite informal. In later sections proofs will usually be informal. The proof of 1.4.1 in a more formal style is as follows:

$$\begin{aligned} \sim(p \vee q) &= \sim(\sim\sim p \vee \sim\sim q), && \text{1.3.2(f)}, \\ &= \sim p \wedge \sim q, && \textit{Sub}\,(p : \sim p, q : \sim q), \quad \text{1.3.1(a)}. \end{aligned}$$

Similarly, 1.4.2 is proved by

$$\begin{aligned} \sim(p \wedge q) &= \sim[\sim(\sim p \vee \sim q)], && \text{1.3.1(a)}, \\ &= \sim p \vee \sim q, && \textit{Sub}\,(p : \sim p \vee \sim q), \quad \text{1.3.2(f)}. \end{aligned}$$

In addition to their use in establishing the relationships between $\vee$ and $\wedge$ for later use, De Morgan's laws are important in their own right. Observe that 1.4.1 states that the negation of *p or q* is *not-p and not-q*, and that 1.4.2 states that the negation of *p and q* is *not-p or not-q*. Hence the negation of "Red and yellow are primary colors" is "Red or yellow is not a primary color."

Now we use De Morgan's laws to prove the following theorems which illustrate the properties of $\wedge$.

1.4.3 $p \wedge q = q \wedge p$ (commutativity of $\wedge$)

1.4.4 $p \wedge (q \wedge r) = (p \wedge q) \wedge r$ (associativity of $\wedge$)

1.4.5 $p \wedge (q \vee r) = (p \wedge q) \vee (p \wedge r)$

(distributivity of $\wedge$ over $\vee$)

1.4.6 $p \wedge p = p$

Proofs will be given for 1.4.3 and 1.4.6, and the proofs of the other theorems will be left as exercises.

To prove 1.4.3, we write

$$\begin{aligned} p \wedge q &= \sim(\sim p \vee \sim q) && \text{1.3.1(a)} \\ &= \sim(\sim q \vee \sim p) && \textit{Sub}\,(p : \sim p, q : \sim q), \quad \text{1.3.2(b)} \\ &= q \wedge p && \textit{Sub}\,(p : q, q : p). \end{aligned}$$

Theorem 1.4.6 can be proved by

$$\begin{aligned} p \wedge p &= \sim(\sim p \vee \sim p) && \textit{Sub}\,(q : p),\quad 1.3.1(\text{a}) \\ &= \sim(\sim p) && \textit{Sub}\,(p : \sim p),\quad 1.3.2(\text{e}) \\ &= p && 1.3.2(\text{f}). \end{aligned}$$

In view of these theorems involving $\wedge$, it should be apparent that it is possible to take $\wedge$, $\sim$, and $=$ as undefined and to state axioms in terms of $\wedge$ rather than $\vee$ and still develop the same theorems. Exercises at the end of this section suggest alternative methods of defining the terms and developing the theory.

In most mathematics the theorems in which one is interested are "if..., then..." or "if and only if" statements. Hence it behooves us to pay special attention to obtaining theorems which provide equivalent ways of formulating $p \Rightarrow q$. The following will be useful later.

1.4.7 $p \Rightarrow q = \sim q \Rightarrow \sim p$ (contrapositive).

1.4.8 $p \Rightarrow q = p \wedge \sim q \Rightarrow \sim p.$

1.4.9 $p \Rightarrow q = p \wedge \sim q \Rightarrow q.$

1.4.10 $p \Rightarrow q = p \wedge \sim q \Rightarrow r \wedge \sim r.$

A proof will be given for 1.4.7, and the remaining proofs will be left as exercises.

$$\begin{aligned} &(1)\quad & p \Rightarrow q &= \sim p \vee q && 1.3.1(\text{b}) \\ &(2) & &= q \vee \sim p && \textit{Sub}\,(p : \sim p),\quad 1.3.2(\text{b}) \\ &(3) & &= \sim\sim q \vee \sim p && \textit{Sub}\,(p : q),\quad 1.3.2(\text{f}) \\ &(4) & &= \sim q \Rightarrow \sim p && \textit{Sub}\,(p : \sim q, q : \sim p),\quad 1.3.1(\text{b}). \end{aligned}$$

When a proof is based on one of the above theorems, the proof is said to be an *indirect proof.* Next we give an example of an indirect proof. We prove the following theorem: *If x and y are integers and xy is odd, then x and y are both odd.* Suppose that xy is odd and that x or y is even, say $x = 2m$ for some integer m. Then $xy = (2m)y = 2(my)$. Hence xy is even, a contradiction. Hence when xy is odd, x and y are odd.

A convention. To simplify the writing of proofs, let us agree to permit some abuse of language. For instance, in step (3) above, it should be clear what substitution is made in 1.3.2(f); hence it would suffice to give the following as a justification of step (3): *Sub* 1.3.2(f). Similarly, steps (2) and (4) could be made more brief. This convention will save considerable time in the writing of proofs.

After these rudimentary theorems have been established, the variety and number of theorems which are possible are limited only by one's imagination. The theorems we have proved here are sufficient as background to prove almost any theorem one will need in his future mathematical endeavors. A large selection of useful theorems is listed here for future reference. The proofs are left as exercises.

1.4.11 $p \vee q \Rightarrow q \vee p$.

1.4.12 $p \wedge q \Rightarrow q \wedge p$.

1.4.13 $(p \Leftrightarrow q) \Rightarrow (q \Leftrightarrow p)$.

1.4.14 $(p \Leftrightarrow q) \Rightarrow (\sim p \Leftrightarrow \sim q)$.

1.4.15 $(p \vee q) \wedge \sim p = q$.

1.4.16 $[p \Rightarrow (q \Rightarrow r)] = [(p \wedge q) \Rightarrow r]$.

1.4.17 $[p \Rightarrow (q \Rightarrow r)] = [q \Rightarrow (p \Rightarrow r)]$.

1.4.18 $[(p \wedge q) \Rightarrow p] = [q \Rightarrow (p \vee \sim p)]$.

1.4.19 $[(p \Rightarrow q) \wedge (p \Rightarrow r)] = [p \Rightarrow (q \wedge r)]$.

1.4.20 $[p \wedge (\sim p \vee q)] \Rightarrow q$.

1.4.21 $[p \wedge (\sim q \Rightarrow \sim p)] \Rightarrow q$.

1.4.22 $[(p \Rightarrow q) \wedge (q \Rightarrow r)] \Rightarrow [p \Rightarrow r]$.

1.4.23 $[(p \Rightarrow q) \wedge (r \Rightarrow s)] \Rightarrow [(p \vee r) \Rightarrow (q \vee s)]$.

1.4.24 $[(p \Rightarrow r) \wedge (q \Rightarrow r)] \Rightarrow [(p \vee q) \Rightarrow r]$.

1.4.25 $(p \Leftrightarrow q) \Rightarrow [(p \wedge r) \Leftrightarrow (q \wedge r)]$.

1.4.26 $[(p \Leftrightarrow q) \wedge (r \Leftrightarrow s)] \Rightarrow [(p \vee r) \Leftrightarrow (q \vee s)]$.

1.4.27 $\{p \wedge [(p \wedge \sim p) \Rightarrow \sim p]\} \Rightarrow q$.

1.4.28 $[(p \Rightarrow q) \wedge (p \vee r)] \Rightarrow (q \vee r)$.

1.4.29 Exercises

1. Form negations of the statements in 1.2.1, Exercises 1 and 2. Express the negations in contemporary English.
2. Write out proofs for 1.4.4 and 1.4.5.
3. Write out proofs for 1.4.8, 1.4.9, and 1.4.10.
4. Write out proofs for 1.4.11 through 1.4.28.
5. Express the following logical formulas in terms of $\sim$ and $\wedge$ only.

   a) $\sim p \Rightarrow q$. b) $p \wedge [\sim(q \vee r)]$. c) $p \vee (q \sim p)$.
   d) $(p \Rightarrow q) \vee (\sim p \Rightarrow \sim q)$. e) $(p \Rightarrow q) \Rightarrow r$.
   f) $[p \Rightarrow (q \veebar p)] \Rightarrow q$.

6. Investigate the properties of $\veebar$. Prove all laws involving it. In particular, answer these questions:

   a) Is $\veebar$ commutative or associative?
   b) Is $\veebar$ distributive over $\wedge$, $\vee$, or $\Rightarrow$?
   c) Can $\veebar$ be expressed in terms of $\sim$ and $\wedge$?
   d) Is this a theorem: $p \veebar q = p \Leftrightarrow \sim q$?

7. Take $\Rightarrow$ and $\sim$ as undefined and define the other connectives in terms of them. Express axioms in terms of $\Rightarrow$, $\sim$, and $=$ which are equal to 1.3.2(a) through (f).
8. Define a new connective / by saying that p/q is false when p and q are true and p/q is true otherwise.

   a) Define all the other connectives in terms of /.
   b) Prove six laws involving /.
   c) What would be the objection to using / as the undefined term, and then developing the logical system?

9. Write indirect proofs for the following statements. Use the informal literary style of the example above.

   a) An integer x is odd if and only if x^2 is odd.
   b) For integers x and y, xy is even if and only if x or y is even.
   c) If I do homework regularly, I will pass this course.

1.5 QUANTIFIERS

Phrases which convey the idea of quantity are called *quantifiers*. For instance, there are propositions which are true for *all* choices of their variables; some which are true for *some* values of their variables; and some which are true for *no* choices of their variables. In this section we want to make the concept of quantifiers exact and develop a notation for expressing quantifiers in symbolic language.

The quantifier *all* is called the *universal quantifier*. The symbol $\forall x$ will be used to mean "for all x" or "for each x." Hence if p is a proposition which depends only on x, to say "p is a theorem" is the same as saying "$\forall x\, p$"; that is, "for all x, p." Hence $\forall p\, p \wedge p = p$. The commas which help make English sentences more readable are usually omitted in symbolic sentences.

The second quantifier is *some*. The symbol $\exists x$ will be used to mean "for some x" or "there exists an x." This quantifier is called the *existential quantifier*. The word *some* is used in mathematics to mean *at least one*. Examples of the use of the existential quantifier are $\exists x\, x^2 = 4$ and $\exists p\, q \Rightarrow p$.

A third expression *none* or *no* could also be considered as a quantifier, but the idea of none can be expressed in terms of the existential quantifier and a negation. For instance, "No horses are pink" can be expressed as "There does not exist a horse which is pink." If p is true for no values of x, we can write $\sim\exists x\, p$. Consequently, we will not consider none as a separate quantifier although some authors do.

These quantifiers will be used in a much wider context than in symbolic logic. Hence if $f(x)$ is any expression involving an x, we can write $\forall x\, f(x)$ and $\exists x\, f(x)$ to express the quantification of the statement $f(x)$. In particular, $f(x)$ may be a logical formula where x is one of the propositions involved. Hence the remainder of our discussion of quantifiers will be done with $f(x)$, but please realize that this includes the application to symbolic logic as a special case.

Our intuitive feelings concerning the universal and existential quantifiers tell us that a universal statement $\forall x\, f(x)$ is false if there is an x such that $f(x)$ is not true; that is, $\sim\forall x\, f(x) = \exists x \sim f(x)$. Similarly, the negation of an existential statement is a universal statement of a negation. Hence for a formal definition of the quantifiers $\forall x\, f(x)$ and $\exists x\, f(x)$, we take $\exists x\, f(x)$ as undefined and make the following definition.

1.5.1 Definition. $\forall x\, f(x) = \sim\exists x \sim f(x)$.

From the definition one can easily verify the following.

1.5.2 Theorem

a) $\exists x\, f(x) = \sim\forall x \sim f(x)$.
b) $\sim\forall x\, f(x) = \exists x \sim f(x)$.
c) $\sim\exists x\, f(x) = \forall x \sim f(x)$.

This definition and theorem must be kept in mind when forming the negations of sentences which involve quantifiers. For instance, the negation of "All horses are pink" is "There exists a horse which is not pink." Practice in negating sentences involving quantifiers is provided at the end of the section.

The most useful applications of quantifiers occur when there is more than one variable to be governed by a quantifier. There are eight ways in which the quantifiers can be combined when two variables are present:

1) $\forall y \forall x \cdots =$ For every y, for every $x \cdots$
2) $\forall x \forall y \cdots =$ For every x, for every $y \cdots$
3) $\forall x \exists y \cdots =$ For every x there exists a y such that $\cdots$
4) $\forall y \exists x \cdots =$ For every y there exists an x such that $\cdots$
5) $\exists y \forall x \cdots =$ There exists a y such that for every $x \cdots$
6) $\exists x \forall y \cdots =$ There exists an x such that for every $y \cdots$
7) $\exists x \exists y \cdots =$ There exists an x such that there exists a y such that $\cdots$
8) $\exists y \exists x \cdots =$ There exists a y such that there exists an x such that $\cdots$

Again appealing to our intuition, we can see that (1) = (2) and (7) = (8). In fact, we state this fact as an axiom.

1.5.3 Axiom

a) $\exists x \exists y\, f(x, y) = \exists y \exists x\, f(x, y)$.
b) $\forall x \forall y\, f(x, y) = \forall y \forall x\, f(x, y)$.

It should be clear to the reader that in an expression involving both quantifiers, it is not possible to interchange them without changing the meaning. For instance, compare the following sentences. "There exists a horse such that for each cart, the horse pulls the cart." "For each cart, there exists a horse such that the horse pulls the cart." These can be represented symbolically by $\exists x \forall y\, f(x, y)$ and $\forall y \exists x\, f(x, y)$, respectively. In the first sentence one horse pulls all the carts, whereas in the second sentence there is one horse per cart. The reader should construct other examples which show how changes in the order of quantifiers change the meaning of a sentence.

By applying 1.5.1 and 1.5.2 one can readily establish the following theorem dealing with the negation of sentences involving multiple quantification.

1.5.4 Theorem

a) $\sim[\forall x \forall y\, f(x, y)] = \exists x \exists y \sim f(x, y)$.
b) $\sim[\exists x \exists y\, f(x, y)] = \forall x \forall y \sim f(x, y)$.
c) $\sim[\forall x \exists y\, f(x, y)] = \exists x \forall y \sim f(x, y)$.
d) $\sim[\exists x \forall y\, f(x, y)] = \forall x \exists y \sim f(x, y)$.

1.5.5 Exercises

1. Express each of the following sentences in symbolic form and then give its negation. Express the negation in acceptable English.

   a) All men are mortal.
   b) Some men are rich but unhappy.
   c) Some humans are not mortal.
   d) An elephant never forgets.
   e) Not all professors are stuffed shirts.
   f) No student who is intelligent will fail.
   g) No student will pass unless he learns the material covered in a course.
   h) Not every boy will be happy if some of the girls are eliminated.

2. Follow the instructions for Exercise 1.

   a) For all x, there exists y such that $xy = 1$.
   b) There exists a v such that for all u, $uv = u$.
   c) Not all boys like all girls.
   d) Some girls like all boys.
   e) If all people studied hard all the time, then some of the unsolved problems would be solved.
   f) For each boy and each girl, there is at least one correct mate.
   g) Some people know all the answers.

3. Produce examples to show that dissimilar quantifiers are not commutative; i.e., show by examples that (3) through (6) in the list of double quantifiers above are not the same.

1.6 VALID ARGUMENTS

One of the principal reasons why a mathematician studies symbolic logic is so that he can analyze the arguments which are given as proofs of theorems. An *argument* is a series of statements starting from the hypothesis (or given information) and ending with the conclusion. An argument is said to be *valid* if the hypothesis implies the conclusion, i.e., if the conjunction of the statements of the hypothesis implies the conclusion. The separate statements in the hypothesis are called *premises* or *hypotheses*. Thus an argument is valid if whenever the premises are all true, the conclusion is true.

It must be pointed out, however, that the validity of an argument does not depend on the truth of the conclusion. It is quite possible (and we will see examples of it) for an argument to be valid and the conclusion to be false. Of course, in this case at least one of the premises must be false.

As an illustration of a valid argument, consider the following:

"The wind will blow or it will rain or it will be a nice day. If it rains, I will go to the show, and if it is a nice day, I will go to the beach. If I go to the show or to the beach, then either the wind will blow or it will be a nice day. The wind will not blow. Therefore it will be a nice day." (w, r, n, s, b)

Here the (w, r, n, s, b) after the argument means that w, r, n, s, and b will be used for the propositions in the argument in the order listed; hence the third proposition introduced—namely, "it will be a nice day"—will be represented by n.

Before checking the validity of this argument, let us write it symbolically:

$$\begin{array}{l} w \vee (r \vee n) \\ (r \Rightarrow s) \wedge (n \Rightarrow b) \\ (s \vee b) \Rightarrow (w \vee n) \\ \sim w \\ \hline \therefore\ n \end{array}$$

Here the premises are stated, a line is drawn, and then the conclusion is stated. The symbol $\therefore$ stands for *therefore.*

The following proof verifies that this argument is indeed valid.

(1)	$w \vee (r \vee n)$	premise
(2)	$(r \Rightarrow s) \wedge (n \Rightarrow b)$	premise
(3)	$(s \vee b) \Rightarrow (w \vee n)$	premise
(4)	$\sim w$	premise
(5)	$r \vee n$	*Sub* 1.4.15
(6)	$(r \vee n) \Rightarrow (s \vee b)$	*Inf* 1.4.23, (2)
(7)	$s \vee b$	*Inf* (6), (5)
(8)	$w \vee n$	*Inf* (3), (7)
(9)	n	*Sub* 1.4.15

Thus the conclusion follows from the premises by logical steps which were developed from the definitions, axioms, and rules of proofs stated in the previous sections.

As previously noted, it is possible for an argument to be valid without the conclusion being true. For instance, consider the following argument:

> "If she is a natural blonde, then her pigmentation is light. Persons with light pigmentation have sensitive feelings. Therefore natural blondes have sensitive feelings." (b, p, f)

Of course, the second premise is false. Hence even though the argument is valid, the conclusion may be false. On the other hand, the next argument has a true conclusion and the argument is valid even though the premises are false.

"If it rains, the river will flood. If it floods, the ground will get wet. Therefore, if it rains, the ground will get wet." (r, f, g)

It is our task to analyze arguments to determine whether or not they are valid. As the above examples show, this does not depend on the truth of the conclusion or the premises. We are interested only in the question, "Is the conclusion implied by the conjunction of the premises?" and nothing more.

It is possible to use the methods of indirect proof (1.4.7 through 1.4.10) to test the validity of an argument. Consider the following argument:

"The tree will live if and only if it gets tender loving care. If the tree gets tender loving care, then the weeds will not be removed from the lawn. The weeds are removed from the lawn. Therefore the tree will die." (t, c, r)

Symbolically this argument is given by

$$\begin{array}{l} t \Leftrightarrow c \\ c \Rightarrow \sim r \\ r \\ \hline \therefore \ \sim t \end{array}$$

The steps in the verification of its validity follow. The methods of indirect proof are employed. Supply the missing reasons.

(1) $t \Leftrightarrow c$	(5) c
(2) $c \Rightarrow \sim r$	(6) $\sim r$
(3) r	(7) $r \wedge \sim r$
(4) t supposition	(8) $\sim t$ *Sub* 1.4.10

We have discussed the procedure for establishing that a certain argument is valid. But what if the argument is invalid? Surely we cannot conclude that an argument is invalid simply because we cannot produce a proof for it. Of course not! Let us re-examine what it means for an argument to be valid. An argument is valid if the conjunction of its premises implies the conclusion, i.e., if the conditional $p \Rightarrow q$, where p is the conjunction of the premises and q is the conclusion, is a theorem. Next let us ask, "Under what conditions will $p \Rightarrow q$ not be a theorem?" Of course, the answer is that $p \Rightarrow q$ is false when p is true and q is false. Considering that in this case p represents the conjunction of the premises, we discover that an argument is invalid if it is possible for all premises to be true and yet for the conclusion to be false.

One method of selecting conditions under which an invalid argument has all premises true and yet has a false conclusion is the construction of truth tables. Consider the following argument:

> "The team will win if and only if Bill is able to play. Bill is able to play or his leg is broken. His leg is not broken. Therefore the team will lose." (w, p, b)

Symbolically this argument is

$$\begin{array}{l} w \Leftrightarrow p \\ p \vee b \\ \sim b \\ \hline \therefore\ \sim w \end{array}$$

This argument is valid if

$$[(w \Leftrightarrow p) \wedge (p \vee b) \wedge \sim b] \Rightarrow \sim w \tag{1}$$

is a theorem. Suspecting that this argument is invalid, we construct a truth table:

w	p	b	$w \Leftrightarrow p$	$p \vee b$	$\sim b$	$\sim w$	(1)
T	T	T	T	T	F	F	T
T	T	F	T	T	T	F	F
T	F	T	F	T	F	F	T
T	F	F	F	F	T	F	T
F	T	T	F	T	F	T	T
F	T	F	F	T	T	T	T
F	F	T	T	T	F	T	T
F	F	F	T	F	T	T	T

Consequently, when w and p are true and b is false, the conclusion does not follow from the conjunction of the premises. This proves that the argument is invalid.

In attempting to determine the validity or invalidity of a given argument, the student has a choice of whether he constructs the truth table first or attempts to write a proof first. The alternative chosen should depend upon the student's intuitive appraisal of the argument; if he thinks it is valid, he should try to write a proof for it, and if he thinks it is invalid, he should construct the truth table.

1.6.1 Exercises

1. Test the validity of the following arguments.

   a) If I study, I learn. If I don't study, I have a good time. Therefore, either I learn or I have a good time. (s, l, g)

   b) My books are in the desk or on the floor. They are not on the desk. Therefore, they are on the floor. (d, f)

   c) It is not true that mathematicians are sadistic or lecherous. Therefore, mathematicians are not sadistic. (s, l)

   d) If the chain is pulled, the bell will ring. Hence if the chain is pulled, either the bell will ring or the chain is broken. (p, r, b)

   e) It was either stupidity or carelessness that caused the error. Had it been stupidity, I would have received no credit. But I received partial credit. Therefore it must have been carelessness. (s, c, r)

   f) More sections will be added only if the enrollment increases. If the enrollment increases, the number of instructors will have to increase also. More sections will be added. Thus the number of instructors will increase. (s, e, i)

   g) If Mary dates John, she will be elected football queen. If Mary dates Jim, she will be elected basketball queen. If she is elected either queen, she will definitely be chosen as school sweetheart. She is not chosen as school sweetheart. Thus she did not date John or Jim. (o, f, i, b, s)

   h) Either the average students like the class or the good students are bored with it, but if the average students like it, then the quantity and quality of the material presented must be restricted. If the good students are bored, then the quality of the material is restricted only if the average students like the course. Hence if the average students like the course, the quantity of the material presented must be restricted. (a, g, t, l)

   i) The session is successful only if the participants react when they return home. The session is successful only if the students see improvement in instruction when they go to class. The session is successful and either the participants react or the students see improvement in instruction. The participants do not return home. Therefore the students go to class. (s, r, h, i, c)

   j) If you do not marry for money, you will be happy, and if you are insincere, you will marry for money. If you are happy, you are insincere. If you marry for money, Martha is not the girl for you. Therefore Martha is the girl for you. (m, h, i, g)

   k) If she goes, Bob will be happy and Steve will make money. If it does not rain, then she will go if Steve makes money. Bob is not happy and the building is open. Therefore if the building is open and it does not rain, then Bob will be happy. (g, h, m, r, b)

   l) Had he made a high score, he would have been proud, and had he made a low score, he would have been disgusted. Had he been either proud or disgusted, he would have been kind. He was not kind. Therefore he did not make a high score or a low one. (h, p, l, d, k)

m) If I go out for football, then if I practice conscientiously, I will make the team. If I make the team and the team wins the championship, I will be surprised. If the team loses one game, then even though the team wins the championship, I will not be surprised. Therefore if I go out for football, then if I practice conscientiously, then the team loses one game. (f, p, t, c, s, l)

n) If Bob goes out for football, then Jim and Tom will go out for basketball. If either Jim or Tom goes out for basketball, then the football team will do poorly. Therefore if Bob goes out for football, the football team will do poorly. (B, J, T, f)

o) If the new building is built here, the street must be closed. If the new building is not built here, the street will not be closed. Either the new building is built here or the chemistry building will have to be torn down. Therefore the mathematics building will not be done by fall. (b, s, c, m)

2. Test the validity of the arguments in Exercise 1 using methods of indirect proof.

3. Test the validity of the following arguments.

a)
$$\begin{array}{l} q \Rightarrow p \\ \underline{\sim p \Rightarrow \sim s} \\ \therefore\ s \Rightarrow q \end{array}$$

b)
$$\begin{array}{l} p \Rightarrow q \\ \underline{\sim r \Rightarrow \sim q} \\ \therefore\ \sim r \Rightarrow \sim p \end{array}$$

c)
$$\begin{array}{l} p \vee q \\ \sim q \Rightarrow r \\ \sim p \vee \sim r \\ \underline{\sim (p \wedge q)} \\ \therefore\ \sim p \end{array}$$

d)
$$\begin{array}{l} p \Rightarrow (q \wedge r) \\ s \Rightarrow (p \Rightarrow r) \\ \underline{t \vee \sim q} \\ \therefore\ (s \wedge t) \Rightarrow q \end{array}$$

e)
$$\begin{array}{l} (p \Rightarrow q) \wedge (p \Rightarrow r) \\ (q \vee r) \Rightarrow s \\ \underline{s} \\ \therefore\ p \end{array}$$

f)
$$\begin{array}{l} (p \wedge q) \Rightarrow r \\ s \Rightarrow p \\ \sim s \Rightarrow (\sim u \Rightarrow w) \\ \underline{(r \wedge \sim w) \Rightarrow p} \\ \therefore\ w \Rightarrow p \end{array}$$

4. Test the validity of the arguments in Exercise 3 using methods of indirect proof.

1.7 APPLICATION OF SYMBOLIC LOGIC TO SWITCHING CIRCUITS

The symbolic logic which we have developed has many applications to subjects other than mathematics. For, after all, symbolic logic is simply a formalization of clear thinking. A particular example where the logical formulas we have established play a central role is in the theory of switching circuits. The simplicity of this application of symbolic logic will be a pleasant change of pace after the section on validity of arguments. However, this section is not essential for understanding the later sections and consequently it may be omitted.

No attempt will be made to formalize the concepts and definitions involved here. An appeal is made to the fundamental theory of electricity, which (we hope) is common knowledge. This common knowledge does not go beyond understanding why a switch turns a light on or off.

A *network* is an arrangement of wires and switches which connect to an electrical power supply. Each switch will have an "on" and an "off" position; current flows when the switch is in the "on" position and no current flows when the switch is in the "off" position. We want to consider two problems involving a network. First, given a network and information concerning the position of the switches in the network, determine whether or not current will flow. Second, given a physical situation and a stipulated lighting plan, design a network which will accomplish the desired outcome.

A switch P will be indicated in a diagram by

(1)

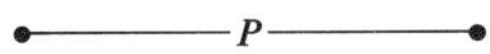

where the large dots at the ends of the network are assumed to be power terminals. If P is on, current will flow, and if P is off, no current will flow.

There are two basic ways in which more than one switch can be combined. The first is to connect switches P and Q *in series*. This is illustrated by the diagram

(2)

Current will flow in this network only if P is on and Q is on. Hence if we let p be the statement "P is on" and let q be "Q is on," then current flows through the network which has P and Q in series if and only if $p \wedge q$ is true.

The second configuration involving two switches P and Q is obtained when P and Q are connected *in parallel*:

(3)

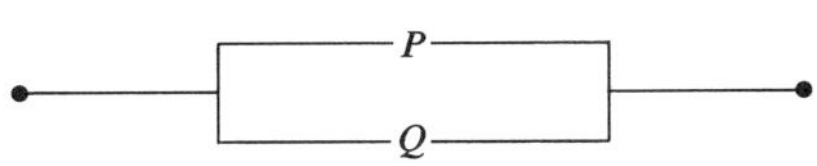

Current will flow in this network only when P is on or Q is on. Thus with p and q as above, current flows in (3) if and only if $p \vee q$ is true.

More complex networks can be generated by combining various combinations of the configurations (1), (2), and (3). For instance,

(4)

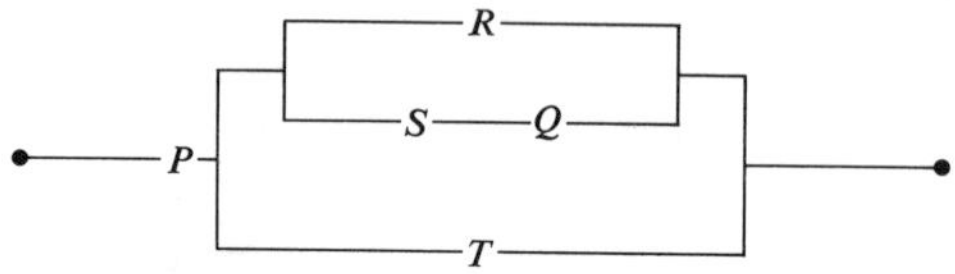

is a network involving five switches, P, Q, R, S, and T. If p, q, r, s, and t are statements saying that the respective switches are on, then current flows in (4) if and only if

$$p \wedge [t \vee (r \vee (s \wedge q))]$$

is true.

It is possible for a given switch to occur more than once in a network. This happens when switches are wired so that current flows through several wires when the switch is on. One way this is done is to connect several switches so that they open and close simultaneously. Moreover, given a switch P, it is possible to have a switch P' which is on when P is off and which is off when P is on. This is accomplished by a "single pull–double throw" switch. If p is the statement "P is on," then $\sim p$ is the statement "P' is on." Consequently, current flows in the network

(5)

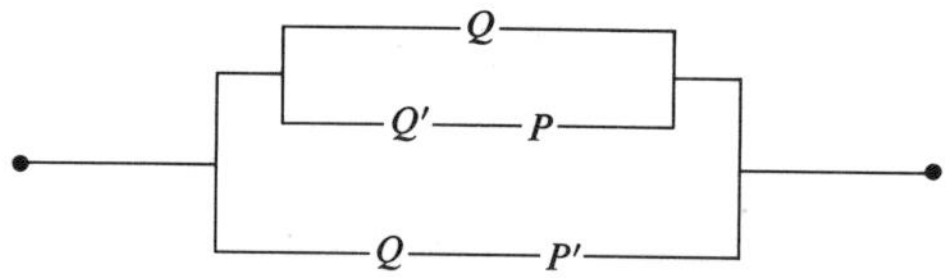

when $(q \wedge \sim p) \vee [q \vee (\sim q \wedge p)]$ is true.

Now the logical identities which we proved in Section 1.4 can be applied to simplify this statement. Hence current flows through (5) if $p \vee q$ is true. Hence (5) is equivalent to (3).

To construct a network which will perform a prescribed function, one must first express the prescribed function in terms of logical symbols and then construct a circuit which will fit the logical symbols. For instance, consider the following situation. A house has a T-shaped hallway:

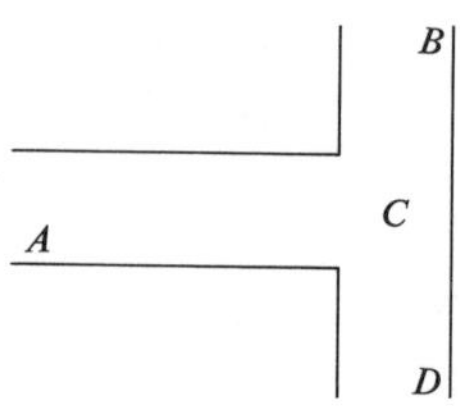

We want a circuit such that there are switches at A, B, and D and a light at C and such that the light can be turned on or off at any switch; e.g., it is possible to turn the light on at A and off at B.

Observe that the light will be on when an odd number of switches are on and the light will be off when an even number of switches are on. If a, b, and d are statements saying that A, B, and D are on, respectively, then current must flow when

$$(a \wedge \sim b \wedge \sim d) \vee (\sim a \wedge b \wedge \sim d) \vee (\sim a \wedge \sim b \wedge d) \vee (a \wedge b \wedge d)$$

is true. Thus a circuit which will perform the required lighting is given by

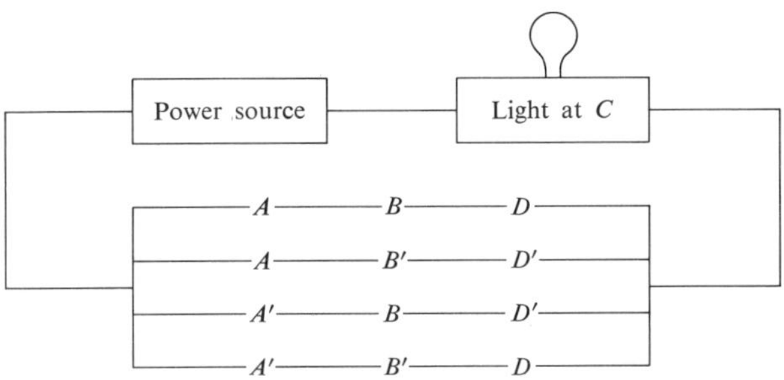

1.7.1 Exercises

1. For each of the following circuits, design (if possible) a simpler circuit in which the current flow is the same as in the original.

(a)

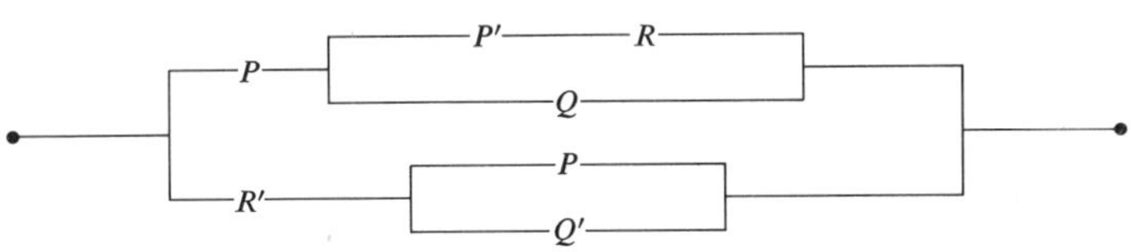

(b)

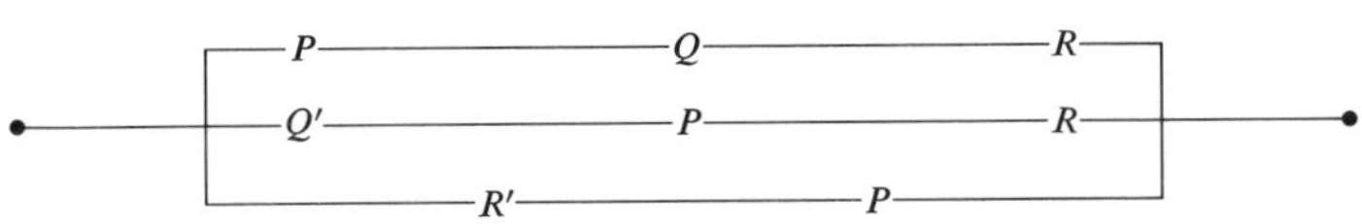

(c)

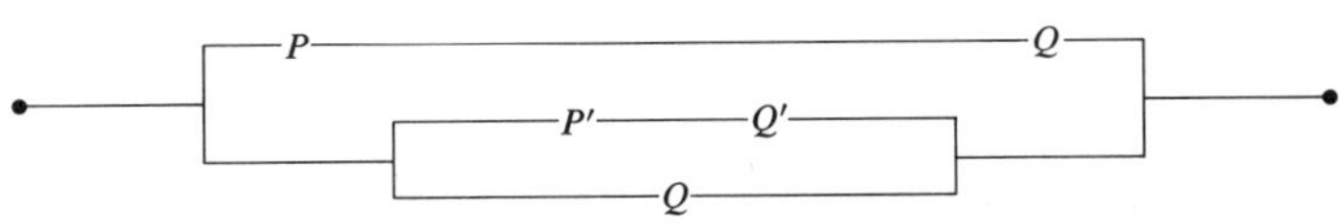

2. Construct a circuit in which the current never flows. Can you characterize, in general, circuits in which the current never flows?

3. A three-man committee wishes to use an electrical system which can be used to record a simple majority vote. Construct a circuit so that each committee member has a switch which he pushes to vote "yes" and such that a bell rings when a majority of the committee votes "yes."
4. To Exercise 3, add the requirement that a light goes on when the majority votes "no."
5. Design a network of switches and a light which will provide a game similar to matching coins. There are two players each of which has a switch that he turns on or off. If both players do the same, the first player wins. Otherwise, the second player wins. Design the circuit such that a light goes on when the first player wins.
6. Design a circuit similar to the one in Exercise 3 which will record a majority vote for a five-man committee in which the chairman has a veto; i.e., if the chairman votes "no," the result is "no."
7. Using as a model the three-switch circuit worked out above, design a similar circuit of four switches, each of which will operate a single light.
8. Design a circuit for giving a four-question true or false exam. The examiner will indicate the correct answers to four questions by setting four switches. The student will indicate his answer, by setting four other switches. Each correct answer turns on a separate white light. If no answers are correct, a red light comes on.

ADDITIONAL REFERENCES FOR CHAPTER 1

Church, A., *Introduction to Mathematical Logic*, Princeton: Princeton University Press (1956).

Copi, Irving M., *Symbolic Logic*, 3rd ed., New York: Macmillan (1967).

Kemeny, John G., Hazleton Mirkil, J. Laurie Snell, and Gerald Thompson, *Finite Mathematical Structures*, Englewood Cliffs, N.J.: Prentice-Hall (1958).

May, Kenneth O., *Elements of Modern Mathematics*, Reading, Mass.: Addison-Wesley (1959).

Mendelson, Elliott, *Introduction to Mathematical Logic*, Princeton: Van Nostrand (1964).

Suppes, Patrick, *Introduction to Mathematical Logic*, Princeton: Van Nostrand (1957).

Tarski, A., *Introduction to Logic*, 2nd rev. ed., New York: Oxford University Press (1946).

2 sets

2.1 DEFINITIONS

The terms *set* and *element of a set* are undefined terms just as point and line are undefined terms in geometry. The word "set" is synonymous with collection. Each set is associated with certain other objects called *elements,* which are said to be *members of the set.* We usually denote sets by capital Latin letters or script letters and elements by lower-case Latin letters. Suppose A is a set and x is an element of A. We denote this relationship symbolically by "$x \in A$." The expression "$x \in A$" is read "x is a member of A," "x belongs to A," or "x is in A." Observe that the symbol $\in$ must have a set on the right and an element on the left. If y is an element which is not a member of B, we write $y \notin B$. For instance, if I denotes the set of all integers, then $3 \in I$ and $\frac{1}{3} \notin I$.

A set may be specified in two ways. First, the elements of the set may be listed within braces; for example, $C = \{a, b, c, d\}$ is a set having exactly four members, a, b, c, and d. Moreover, if Q is the set of positive integers larger than 9, we can write $Q = \{10, 11, 12, 13, \ldots\}$, where ... means "and so on." An alternative method of specifying a set is to describe the elements of the set by listing their properties. For instance, Q above could have been written as

$$Q = \{x \mid x \text{ is a positive integer and } x > 9\}.$$

Moreover, if P is the set of positive integers, then

$$Q = \{x \mid x \in P \text{ and } x > 9\}.$$

One reads this expression as "Q is the set of all x such that $x \in P$ and $x > 9$."

2.1.1 Definition. *For any two sets A and B,*

$$A \subseteq B \Leftrightarrow \forall x (x \in A \Rightarrow x \in B).$$

If $A \subseteq B$, we say A is a *subset* of B or that *A is contained in B*. Note that every set is a subset of itself. The set which has no members is called the *empty* (or *null*) set and is denoted by $\varnothing$. Note that for any set A, $\varnothing \subseteq A$. If A is a nonempty set which is a subset of a set B and if B contains an element which is not a member of A, then A is said to be a *proper* subset of B; we denote this by $A \subset B$. The symbols $\subseteq$ and $\subset$ are used to show a relationship between two sets and not between an element and a set. With one exception, if a is a member of a set A, we may write $a \in A$ or $\{a\} \subseteq A$, but not $a \subseteq A$. The exception is $\varnothing \in \{\varnothing\}$ and $\varnothing \subseteq \{\varnothing\}$.

2.1.2 Definition. *For any two sets A and B,*

$$A = B \Leftrightarrow A \subseteq B \wedge B \subseteq A.$$

If $A = B$, we say that *A equals B*. Thus to prove that two sets A and B are equal, it suffices to prove that A is a subset of B and B is a subset of A.

There are two operations that combine two sets.

2.1.3 Definition. *The* ***intersection*** *of two sets A and B is the set*

$$A \cap B = \{x \mid x \in A \wedge x \in B\}.$$

Pictorially the intersection of two sets A and B can be represented by

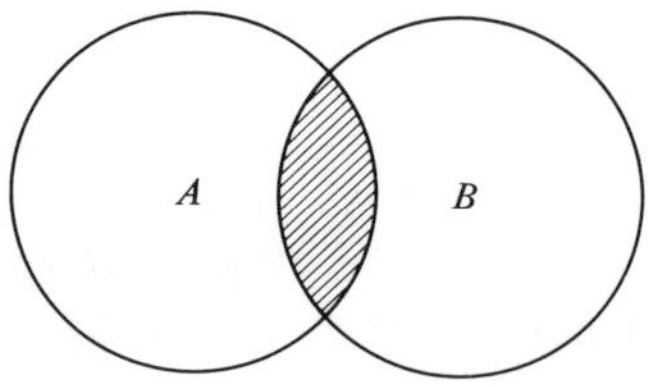

where A is the circle on the left, B that on the right, and $A \cap B$ is the shaded area. If $A = \{1, 2, 3, 4\}$ and $B = \{3, 4, 5\}$, then $A \cap B = \{3, 4\}$. If A is the set of all mathematics majors and B is the set of all coeds, then $A \cap B$ is the set of all coeds majoring in mathematics. For any set A, $A \cap A = A$. For any sets A and B, $A \subseteq B$ if and only if $A \cap B = A$. (Prove it!) If two sets A and B have no elements in common, then their intersection $A \cap B$ is the empty set $\varnothing$. In this case, one says that A and B are *disjoint*.

A second operation used to combine two sets is the *union* of two sets.

2.1.4 Definition. *The **union** of two sets A and B is the set*

$$A \cup B = \{x \mid x \in A \vee x \in B\}.$$

Thus the union of two sets A and B is the set of all x such that x belongs to A or x belongs to B, or both. The union of two sets A and B can be illustrated pictorially by

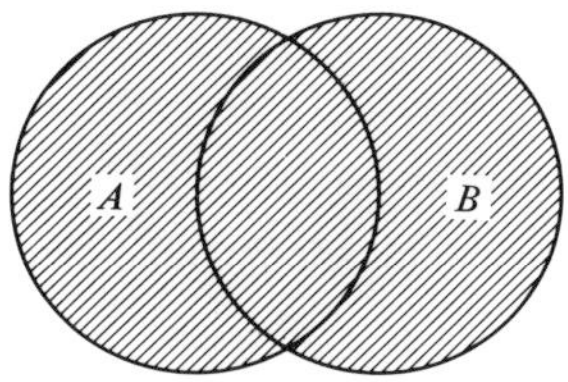

where A is the circle on the left, B that on the right, and $A \cup B$ is the shaded area. If $A = \{1, 2, 3, 4\}$ and $B = \{3, 4, 5\}$, then $A \cup B = \{1, 2, 3, 4, 5\}$. The union of the set A of all mathematics majors and the set B of all coeds is the set $A \cup B$ of all university students who are either mathematics majors or women, or both. For any set A, $A \cup A = A$. For any two sets A and B, $A \subseteq B$ if and only if $A \cup B = B$ (Prove it!)

It is easy to extend the concepts of intersection and union of two sets to the intersection and union of a finite number of sets. For instance, if $X_1, X_2, \ldots, X_n$ are sets, then

$$X_1 \cap X_2 \cap \cdots \cap X_n = \{x \mid x \in X_i \text{ for all } i = 1, 2, \ldots, n\}$$

and

$$X_1 \cup X_2 \cup \cdots \cup X_n = \{x \mid x \in X_i \text{ for some } i = 1, 2, \ldots, n\};$$

i.e., the intersection of the sets $X_1, X_2, \ldots, X_n$ is the set of elements which belong to all of the X_i $(1 \le i \le n)$, and the union of the sets $X_1, X_2, \ldots, X_n$ is the set of elements which belong to at least one of the X_i $(1 \le i \le n)$.

Similarly, if we have a collection of sets $\{X_i \mid i = 1, 2, 3, \ldots\}$ indexed by the set of positive integers, we can form their intersection and union. In this case, the union of the X_i is

$$\bigcup_{i=1}^{\infty} X_i = \{x \mid x \in X_i \text{ for all } i = 1, 2, 3, \ldots\},$$

and the intersection of the X_i is

$$\bigcap_{i=1}^{\infty} X_i = \{x \mid x \in X_i \text{ for some } i = 1, 2, 3, \ldots\}.$$

Let U be a given set. If $X_1, X_2, \ldots, X_n$ are all subsets of U such that

$$U = X_1 \cup X_2 \cup \cdots \cup X_n$$

and

$$X_i \cap X_j = \varnothing \qquad \text{for} \quad i \neq j,$$

then we say that $X_1, X_2, \ldots, X_n$ form a *partition* of U. Thus if $X_1, X_2, \ldots, X_n$ form a partition of U, then every element of U belongs to one of the X_i, but no element of U belongs to more than one of the X_i. A partition of a set U divides U into disjoint subsets such that each member of U belongs to exactly one of the subsets. For instance, the sets $\{1, 5\}$, $\{2, 3, 6\}$, $\{7\}$, and $\{4, 8, 9\}$ form a partition of the set $\{1, 2, 3, 4, 5, 6, 7, 8, 9\}$.

2.1.5 Definition. *For two sets A and B, the* ***Cartesian product*** *(or cross product) $A \times B$ of A and B is the set*

$$A \times B = \{(a, b) \mid a \in A, b \in B\}.$$

The Cartesian product of A and B is the set of all ordered pairs (a, b), where $a \in A$ and $b \in B$. If $a, c \in A$ and $b, d \in B$, then $(a, b) = (c, d)$ if and only if $a = c$ and $b = d$. For $A = \{1, 2\}$ and $B = \{x, y, z\}$, $A \times B = \{(1, x), (1, y), (1, z), (2, x), (2, y), (2, z)\}$. The concept of the cross product of two sets can be extended to the cross product of a finite number of sets. The cross product of the sets $X_1, X_2, \ldots, X_n$ is

$$X_1 \times X_2 \times \cdots \times X_n = \{(x_1, x_2, \ldots, x_n) \mid x_i \in X_i \text{ for } i = 1, 2, \ldots, n\};$$

that is, $X_1 \times X_2 \times \cdots \times X_n$ is the set of all n-tuples $(x_1, x_2, \ldots, x_n)$ where the ith entry is from X_i.

Observe that it is not necessarily true that $A \times B = B \times A$. Give examples to show that these sets may be different. When would $A \times B = B \times A$?

Now that we have defined the major terms used in our set theory, we must turn to the task of proving theorems which will make the concepts useful. In doing this we see immediately the strong dependence of our set theory on symbolic logic.

2.1.6 Theorem. *The operations $\cap$ and $\cup$ are associative; i.e., if A, B, and C are sets, then*

a) $A \cap (B \cap C) = (A \cap B) \cap C$,
b) $A \cup (B \cup C) = (A \cup B) \cup C$.

Proof. We will prove (a); (b) is proved similarly:

$$\begin{aligned} A \cap (B \cap C) &= \{x \mid x \in A \wedge x \in B \cap C\} \\ &= \{x \mid x \in A \wedge (x \in B \wedge x \in C)\} \\ &= \{x \mid (x \in A \wedge x \in B) \wedge x \in C\} \\ &= \{x \mid x \in A \cap B \wedge x \in C\} \\ &= (A \cap B) \cap C. \end{aligned}$$

The third equality follows from 1.4.4.

More theorems on the properties of $\cap$ and $\cup$ are listed in 2.1.7, Exercise 5, as exercises.

2.1.7 Exercises

1. If $A = \{0, 2, 4, 6\}$, $B = \{1, 4, 5, 6, 7\}$, $C = \{1, 3, 4, 8\}$, and $D = \varnothing$, compute the following.

   a) $A \cup B$ b) $B \cap C$ c) $B \cup C$
   d) $A \cap C$ e) $A \cap D$ f) $C \cup D$
   g) $(A \cap B) \cup C$ h) $(B \cup C) \cap A$ i) $A \times B$

2. Observe that the expression $A \cap B \cup C$ is meaningless. Should one find the intersection of A and B, then compute the union of $A \cap B$ and C, or should one find the union of B and C first and then compute the intersection of A and $B \cup C$? The first procedure could be expressed by $(A \cap B) \cup C$; the second by $A \cap (B \cup C)$. Find sets A, B, and C such that $(A \cap B) \cup C$ and $A \cap (B \cup C)$ are not identical.

3. If U is a set and A is a subset of U, $\tilde{A}$ will denote the set of all elements in U which are not in A. $\tilde{A}$ is called the *complement of A in U*. Let $U = \{a, b, c, d, e, f, g, h, i\}$, $A = \{a, b, c, d\}$, $B = \{c, d, e\}$, and $C = \{e, h, i\}$. Compute the following sets.

   a) $\tilde{A}$ b) $\tilde{A} \cap \tilde{B}$ c) $(\widetilde{A \cup B})$
   d) $(\widetilde{A \cap B}) \cap C$ e) $\tilde{\tilde{A}}$ f) $(\widetilde{A \cap B) \cup C}$
   g) $(A \cap B) \cup \tilde{C}$ h) $A \cap \tilde{A}$ i) $B \cup \tilde{B}$

4. Give three different partitions of the set of positive integers.

5. *Prove* the following assertions about the equality of two sets.

   a) $A \cap B = B \cap A$ b) $A \cup B = B \cup A$
   c) $A \cap (B \cup C) = (A \cap B) \cup (A \cap C)$
   d) $A \cup (B \cap C) = (A \cup B) \cap (A \cup C)$
   e) $\widetilde{A \cup B} = \tilde{A} \cap \tilde{B}$ f) $\widetilde{A \cap B} = \tilde{A} \cup \tilde{B}$

6. Does $A \times (B \cup C) = (A \times B) \cup (A \times C)$? What about

$$A \cap (B \times C) = (A \cap B) \times (A \cap C)?$$

   Prove your answers.

7. Consider a set of boys. Twelve boys are thin and tall, sixteen are tall, and twenty are thin. How many boys are in the set altogether?
8. A certain set of girls has the following characteristics. Three are cute, smart, and redheaded; five are smart and redheaded; four are cute and redheaded; ten are redheaded; eight are cute and smart; eighteen are cute; sixteen are smart; and nineteen are neither cute nor smart nor redheaded. How many girls are in this set?
9. Suppose a set A has six elements. How many different ways are there to partition A if we do not allow the empty set to be used? What if A has n elements?

2.2 MAPPINGS

Let us first consider an example. Let C be the set of pages in this book, and let P be the set of positive integers. Then by numbering the pages we are associating with each element of C a certain number, an element of P. This is an example of a mapping from the set C to the set P. Other examples will follow. Right now let us give a precise definition of a mapping.

2.2.1 Definition. *A* ***mapping*** *α from a set A to a set B is a correspondence that associates with each element a of A a unique element b of B. We denote this correspondence by writing $\alpha(a) = b$, and we call b the* ***image of a under α***.

Two shorthand methods of indicating that α is a mapping from a set A to a set B are $\alpha: A \to B$ and $A \xrightarrow{\alpha} B$. It is important to observe that "mapping" is simply another word for "function." The two terms are identical. "Mapping" is most commonly used in algebra, and "function" is usually used in connection with analysis.

There are two words in the definition of mapping that merit examination. One is "each"; in order that α be a mapping defined on the set A, α must be defined for all elements of A. Thus for each $a \in A$, $\alpha(a)$ is some element of B. The second word deserving extra emphasis is "unique." For each $a \in A$, there is only one element of B associated with it; i.e., for each $a \in A$, $\alpha(a)$ can be only one element of B.

Next we define equality of mappings.

2.2.2 Definition. *Two mappings α and β are said to be* ***equal*** *if and only if α and β are defined on the same set A and $\alpha(a) = \beta(a)$ for all $a \in A$.*

Thus mappings α and β defined on a set A are *equal* if they "do the same thing" to every element of A.

2.2.3 Examples

1. Let $A = \{1, 2, 3, 4\}$ and $B = \{u, v, w\}$. Then the association

$$\alpha(1) = u, \qquad \alpha(2) = v, \qquad \alpha(3) = v, \qquad \alpha(4) = w$$

is a mapping from A to B. Note that the fact that 2 and 3 have the same image under α does not contradict the "uniqueness" required in the definition of mapping. The numbers 2 and 3 each have unique images; it simply happens that they both have the same image. But neither of them has more than one image, so α is a mapping.

2. Let A be any set. The correspondence which associates each element of A with itself is clearly a mapping from A to A. We denote this mapping by ι (iota) and call it the *identity mapping* on A. Then $\iota(a) = a$ for all $a \in A$.

3. Let I be the set of all integers and define a correspondence β from I to itself by $\beta(i) = i + 1$ for all $i \in I$. To see that β is a mapping, one must show that for each $i \in I$, $i + 1$ is an element of I and that for each $i \in I$, there is only one element of the form $i + 1$ in I. It is easy to see that both of these conditions hold. Thus β is a mapping.

4. Let I be the set of all integers and define a correspondence ρ from I to itself by $\rho(x) = x^2 - x$ for all $x \in I$. Then ρ is a mapping (check it!).

One can express the definition of a mapping α from a set A to a set B as follows:

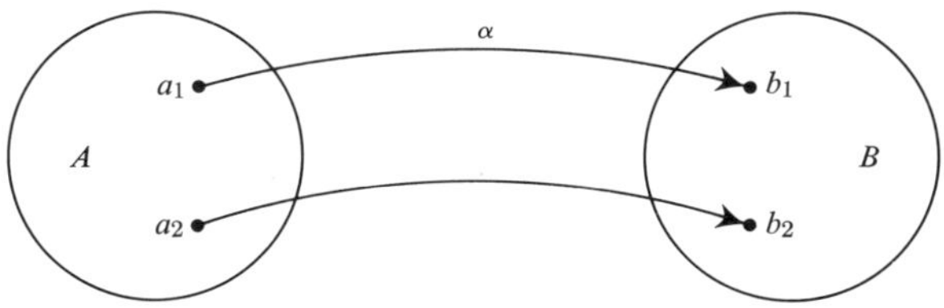

Here the circle on the left denotes A and the one on the right, B. The arrows go from elements of A to their images in B. Thus for α to be a mapping, each element of A must be the initial point of an arrow and no point of A can be the initial point of two arrows. Pictorially the mapping α in Example 1 above would be

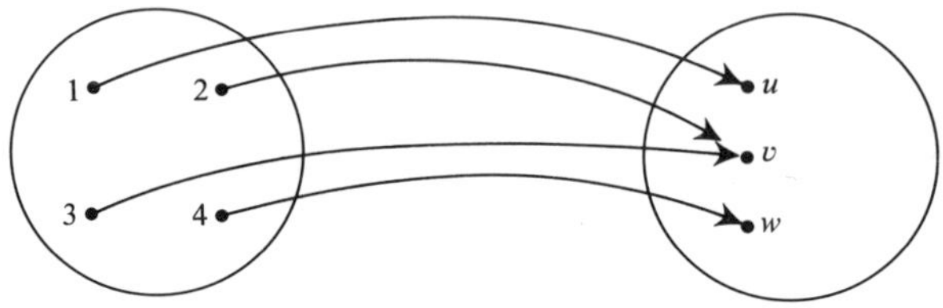

2.2.4 Definition. *A mapping α from a set A to a set B is said to be a mapping of A* ***onto*** *B if every element of B is the image of some element of A.*

Thus if α is a mapping of A *onto* B, then for each $b \in B$, there exists an element $a \in A$ such that $\alpha(a) = b$. In the mapping illustration on p. 31, a mapping $\alpha: A \to B$ would be *onto* B if every element of B was the tip of some arrow. The mapping α in Example 1 is *onto*, since each of the elements u, v, and w of B is an image of some element of A. The identity mapping ι is clearly onto. The mapping β of Example 3 is a mapping from I *onto* I, for given any i in I it is the image of the element $i - 1 \in I$ since $\beta(i - 1) = (i - 1) + 1 = i$. However, the mapping ρ of Example 4 is *not onto*, since the element $3 \in I$ is not the image of any $x \in I$ because there is no integer x such that $x^2 - x = 3$.

2.2.5 Definition. *A mapping α from a set A to a set B is said to be* ***one-to-one*** *if distinct elements of A have distinct images in B.* (Two elements are *distinct* if they are unequal.)

Thus if α is a *one-to-one* mapping from a set A to a set B, then for all elements $a_1, a_2 \in A$ such that $a_1 \neq a_2$, we must have $\alpha(a_1) \neq \alpha(a_2)$. An equivalent way to express this is α is *one-to-one* if $\alpha(a_1) = \alpha(a_2)$ for $a_1, a_2 \in A$ implies that $a_1 = a_2$. Pictorially a mapping is one-to-one if no two arrows have the same tip. The mapping α in Example 1 is not one-to-one, since 2 and 3 have the same image; that is, $\alpha(2) = \alpha(3)$, but $2 \neq 3$. The identity mapping ι is clearly one-to-one. The mapping β of example 3 is one-to-one; for if $i, j \in I$ and $i \neq j$, then $i + 1 \neq j + 1$. Hence $i \neq j$ implies $\beta(i) \neq \beta(j)$. In Example 4, ρ is not one-to-one, since $\rho(0) = \rho(1)$ but $0 \neq 1$ (check!).

To summarize, Example 1 shows a mapping which is onto but not one-to-one; Example 3 one which is onto and one-to-one; Example 4 one which is neither onto nor one-to-one. We leave it as an exercise to find an example of a mapping which is one-to-one but not onto. Thus one can see that the concepts of *onto* mappings and *one-to-one* mappings are completely independent.

Consider the following situation: Let A, B, and C be sets, let α be a mapping from A to B, and let β be a mapping from B to C. Let a be an element of A. Then $\alpha(a)$ is an element of B and consequently it makes sense to ask, "What is the image of $\alpha(a)$ under β?" It would be $\beta(\alpha(a))$, that is, β of $\alpha(a)$. In this way, for each element a of A we can associate an element of C, namely $\beta(\alpha(a))$. Thus α and β can be combined to produce a mapping from A to C. This is the basis of the following definition.

2.2.6 Definition. *If A, B, and C are sets and $\alpha: A \to B$ and $\beta: B \to C$, then the **composition of α and β** is the mapping $\beta \circ \alpha: A \to C$ defined by $(\beta \circ \alpha)(a) = \beta(\alpha(a))$ for all $a \in A$.*

Thus in the situation in this definition, $(\beta \circ \alpha)(a)$ will mean apply α to a first, and then apply β to $\alpha(a)$. Note that in order for the composition $\sigma \circ \rho$ of two mappings σ and ρ to make sense, ρ must have its images in the set on which σ is defined.

Let us illustrate the composition of two mappings by some examples.

2.2.7 Examples

1. Let $A = \{1, 2, 3, 4\}$, $B = \{u, v, w\}$, and $C = \{a, b, c\}$. Define $\alpha: A \to B$ and $\beta: B \to C$ by

$$\alpha(1) = u, \qquad \alpha(2) = v, \qquad \alpha(3) = v, \qquad \alpha(4) = w$$

and

$$\beta(u) = a, \qquad \beta(v) = b, \qquad \beta(w) = b.$$

Let us compute the composition $\beta \circ \alpha$. By definition $(\beta \circ \alpha)(1)$ is the application of α to 1 and then β to the result. Thus

$$(\beta \circ \alpha)(1) = \beta(\alpha(1)) = \beta(u) = a.$$

Thus $(\beta \circ \alpha)(1) = a$. Similarly,

$$(\beta \circ a)(2) = \beta(\alpha(2)) = \beta(v) = b,$$

$$(\beta \circ \alpha)(3) = \beta(\alpha(3)) = \beta(v) = b,$$

and

$$(\beta \circ \alpha)(4) = \beta(\alpha(4)) = \beta(w) = b.$$

Consequently,

$$(\beta \circ \alpha)(1) = a, \qquad (\beta \circ \alpha)(2) = b, \qquad (\beta \circ \alpha)(3) = b, \qquad (\beta \circ \alpha)(4) = b.$$

In this case, $\alpha \circ \beta$ does not make sense.

2. Let A be any set and let ι be the identity mapping on A. If α is any other mapping from A to itself, then $\alpha \circ \iota = \iota \circ \alpha = \alpha$. To see this, let $a \in A$ and observe that $(\alpha \circ \iota)(a) = \alpha(\iota(a)) = \alpha(a)$ and $(\iota \circ \alpha)(a) = \iota(\alpha(a)) = \alpha(a)$.

3. Let I be the set of all integers and let β and ρ be the mappings defined in Examples 3 and 4 of 2.2.3; that is, $\beta: I \to I$ is defined by $\beta(i) = i + 1$ for all $i \in I$ and $\rho: I \to I$ is defined by $\rho(i) = i^2 + i$ for all $i \in I$. Then $\rho \circ \beta$ and $\beta \circ \rho$ both make sense. Moreover,

$$(\rho \circ \beta)(i) = \rho(\beta(i)) = \rho(i + 1) = (i + 1)^2 + i + 1 = i^2 + 3i + 2$$

for all $i \in I$ and

$$(\beta \circ \rho)(i) = \beta(\rho(i)) = \beta(i^2 + i) = (i^2 + i) + 1 = i^2 + i + 1.$$

Hence we see that even if $\rho \circ \beta$ and $\beta \circ \rho$ are both sensible, they may not be the same mapping.

Next we state and prove a lemma which relates an important property of the composition of mappings, namely, that when the necessary compositions make sense, the composition of mappings is associative.

2.2.8 Lemma (Associative Law). *If A, B, C, and D are sets and if $\alpha: A \to B$, $\beta: B \to C$, and $\gamma: C \to D$, then $\gamma \circ (\beta \circ \alpha) = (\gamma \circ \beta) \circ \alpha$.*

Proof. Note first that $\beta \circ \alpha$ and $\gamma \circ \beta$ make sense, and that $\gamma \circ (\beta \circ \alpha)$ and $(\gamma \circ \beta) \circ \alpha$ also make sense. Hence it is feasible to try to prove the lemma.

To prove that the two mappings are equal (see 2.2.2), we must prove for each $a \in A$,

$$(\gamma \circ (\beta \circ \alpha))(a) = ((\gamma \circ \beta) \circ \alpha)(a).$$

But by definition of composition of mappings,

$$(\gamma \circ (\beta \circ \alpha))(a) = \gamma((\beta \circ \alpha)(a)), \qquad \text{and} \qquad (\beta \circ \alpha)(a) = \beta(\alpha(a)).$$

Hence

$$(\gamma \circ (\beta \circ \alpha))(a) = \gamma((\beta \circ \alpha)(a)) = \gamma(\beta(\alpha(a))).$$

On the other hand,

$$((\gamma \circ \beta) \circ \alpha)(a) = (\gamma \circ \beta)(\alpha(a)) = \gamma(\beta(\alpha(a))).$$

Thus the elements $(\gamma \circ (\beta \circ \alpha))(a)$ and $((\gamma \circ \beta) \circ \alpha)(a)$ are indeed equal for each $a \in A$. This proves the lemma.

Now let A and B be sets and let α be a one-to-one mapping from A *onto* B. Then by the "onto-ness" of α, for each $b \in B$ there exists $a \in A$ such that $\alpha(a) = b$. By the "one-to-one–ness" of α, this a is unique. Hence the correspondence associating b to a is a mapping. We denote this mapping by α^{-1} and call it the *inverse* of α. Thus α^{-1} is a mapping from B to A given by $\alpha^{-1}(b) = a$ if $\alpha(a) = b$. Note that in order to define α^{-1} we require that α is one-to-one and onto. The symbol α^{-1} is nonsense if α is not one-to-one and onto.

Note that if α is a one-to-one mapping from A onto B, then α^{-1} is a mapping from B to A. Hence both $\alpha \circ \alpha^{-1}$ and $\alpha^{-1} \circ \alpha$ make sense. Let $a \in A$ and let $\alpha(a) = b$; then

$$(\alpha^{-1} \circ \alpha)(a) = \alpha^{-1}(\alpha(a)) = \alpha^{-1}(b) = a.$$

Thus $\alpha^{-1} \circ \alpha$ is the identity mapping on A. Similarly, $\alpha \circ \alpha^{-1}$ is the identity mapping on B.

2.2.9 Definition. *Let A be any nonempty set. Let $S(A)$ denote the set of all one-to-one mappings of A onto itself.*

The set $S(A)$ will be an important example in the discussion of the mathematical system known as a group (see Chapter 5). For that reason, we list the properties of $S(A)$ now for future reference.

2.2.10 Properties of $S(A)$. *If α, β, and γ are elements of $S(A)$, then*

1) $\alpha \circ \beta$ *is an element of* $S(A)$.

2) $\alpha \circ (\beta \circ \gamma) = (\alpha \circ \beta) \circ \gamma$.

3) There exists an element ι (the identity mapping on A) in $S(A)$ such that $\alpha \circ \iota = \iota \circ \alpha = \alpha$.

4) There exists an element α^{-1} in $S(A)$ such that $\alpha \circ \alpha^{-1} = \alpha^{-1} \circ \alpha = \iota$.

We have verified properties (2), (3), and (4) in the foregoing discussion. Moreover, since α and β are mappings from A to A, $\alpha \circ \beta$ makes sense. That $\alpha \circ \beta$ is in fact one-to-one and onto is the result of 2.2.11, Exercise 5. This exercise definitely should be done before continuing.

2.2.11 Exercises

1. Give an example of a mapping which is one-to-one, but not onto.
2. Let I be the set of all integers. Which of the following mappings from I to I are one-to-one and which are onto? Prove your answer.

   a) $\alpha(i) = i - 4$ b) $\alpha(i) = i^2$
   c) $\alpha(i) = i^2 + 4$ d) $\alpha(i) = i^3$

3. Let R be the set of real numbers. Which of the following mappings from R to R are one-to-one and which are onto? Prove your answer.

   a) $\rho(x) = x + 6$ b) $\rho(x) = x^2 + 2$
   c) $\rho(x) = x^2 - 2$ d) $\rho(x) = x^3 - x + 2$

4. Let P be the set of positive integers and let E be the set of even positive integers. Give an example of a mapping which satisfies the following conditions.

   a) $\alpha: P \to E$, one-to-one and onto.
   b) $\alpha: P \to E$, one-to-one and not onto.
   c) $\alpha: P \to E$, not one-to-one and onto.
   d) $\alpha: P \to E$, not one-to-one and not onto.
   e) $\alpha: E \to P$, one-to-one and onto.
   f) $\alpha: E \to P$, one-to-one and not onto.
   g) $\alpha: E \to P$, not one-to-one and onto.
   h) $\alpha: E \to P$, not one-to-one and not onto.

5. Prove the following. If $\alpha: A \to B$ and $\beta: B \to C$ are mappings, then

   a) $\beta \circ \alpha$ is one-to-one if both α and β are one-to-one, and
   b) $\beta \circ \alpha$ is onto if both α and β are onto.

2.3 CARDINALITY

Some of the sets which we examined in Section 2.1 contained a limited number of elements, whereas others contained an unlimited number of elements. Quite often it is useful to distinguish between these two types of sets. In order to do this, we introduce the concepts of *finite* and *infinite* sets.

2.3.1 Definition. *A set A is in* ***one-to-one correspondence*** *with a set B if there exists a one-to-one mapping from A onto B. If A is in one-to-one correspondence with B, we write $A \approx B$ and say* ***A is equivalent to B.***

For example, $\{1, 4, 6, 9\} \approx \{a, b, c, d\}$ and $\{2, 4, 1\} \approx \{*, 11, Q\}$. Note that equivalent sets are not necessarily the same. What is a one-to-one correspondence which shows that $\{2, 4, 1\} \approx \{*, 11, Q\}$? Three are listed below:

$$\begin{array}{lll} \alpha(2) = *, & \beta(2) = * & \gamma(2) = 11, \\ \alpha(4) = 11, & \beta(4) = Q, & \gamma(4) = Q, \\ \alpha(1) = Q, & \beta(1) = 11, & \gamma(1) = *. \end{array}$$

Are there more?

Also note that $\{1, 2, 3, 4, \ldots\} \approx \{2, 4, 6, 8, \ldots\}$. One one-to-one mapping from $\{1, 2, 3, 4, \ldots\}$ onto $\{2, 4, 6, 8, \ldots\}$ that shows that these sets are equivalent is $\alpha(n) = 2n$. There are many more mappings which will suffice to prove that these sets are equivalent; find several more!

2.3.2 Remark

Let A and B be sets such that $A \approx B$. Then there exists a one-to-one mapping α from A onto B. However, in view of the discussion in the previous section, α^{-1} exists and is a one-to-one mapping from B onto A. Consequently, $B \approx A$.

2.3.3 Definition. *A set A is said to be* ***infinite*** *if A has a proper subset B such that $A \approx B$. Any set which is not infinite is said to be* ***finite.***

The example considered above shows that $\{1, 2, 3, 4, \ldots\}$ is an infinite set since $\{2, 4, 6, 8, \ldots\}$ is a proper subset of $\{1, 2, 3, 4, \ldots\}$, which is equivalent to it. On the other hand, it is easy to see that $\{1, 2, 3, 4\}$ is a finite set. Try to find a proper subset which is equivalent to it and you will see that none exists.

Intuitively a set A is finite if, when counting the elements A, the process comes to an end, and is infinite if the process does not come to an end. Hence a nonempty set is finite if and only if there is a set of the form $\{1, 2, 3, \ldots, n\}$ which is equivalent to it.

2.3.4 Definition. *Let A be a finite, nonempty set. The* ***cardinality of A,*** *written $n(A)$, is the natural number k if and only if $A \approx \{1, 2, 3, \ldots, k\}$. If A is the empty set, we write $n(A) = 0$.*

Next we see that finite sets are characterized by their cardinality.

2.3.5 Theorem. *Any two finite, nonempty sets A and B are equivalent if and only if $n(A) = n(B)$.*

Proof. Note that this is an "if and only if" statement, so by 1.3.1(c) we must prove two conditional statements.

First assume that $A \approx B$ and prove that $n(A) = n(B)$. Let $n(A) = k$. Then there exists a one-to-one mapping α from A onto $\{1, 2, 3, \ldots, k\}$. Also, since $A \approx B$, 2.3.2 shows that there exists a one-to-one mapping β from B onto A. Thus the composition $\alpha \circ \beta$ is a mapping from B to $\{1, 2, 3, \ldots, k\}$ and 2.2.11, Exercise 5, shows that $\alpha \circ \beta$ is one-to-one and onto. Thus $n(B) = k$ so that $n(A) = n(B)$.

Now assume that $n(A) = n(B)$ and prove that $A \approx B$. Let $n(A) = k$. Then there exists a one-to-one mapping β from A onto $\{1, 2, \ldots, k\}$ and a one-to-one mapping α from B onto $\{1, 2, \ldots, k\}$. Then the composition $\alpha^{-1} \circ \beta$ is a one-to-one mapping from A onto B. Thus $A \approx B$ and the theorem is proved.

Let C be a nonempty finite set and let $A_1, A_2, \ldots, A_n$ be a partition of C. Then every element of C belongs to one and only one of the A_i. Hence the number of elements in C is the sum of the number of elements in the A_i; that is,

$$n(C) = n(A) + n(A_2) + \cdots + n(A_n).$$

We use this fact in the proof of the next theorem.

2.3.6 Theorem. *Let A and B be nonempty finite sets. Then*

$$n(A \cup B) = n(A) + n(B) - n(A \cap B).$$

Proof. Note that if

$$C = \{x \mid x \in A \wedge x \notin B\}$$

and

$$D = \{x \mid x \in B \wedge x \notin A\},$$

then the sets $A \cap B$, C, and D form a partition of $A \cup B$. Hence by the remark above,

$$n(A \cup B) = n(A \cap B) + n(C) + n(D).$$

However, $A \cap B$ and C form a partition of A, and $A \cap B$ and D form a partition of B. Hence

$$n(A) = n(A \cap B) + n(C) \qquad \text{and} \qquad n(B) = n(A \cap B) + n(D).$$

Therefore

$$n(C) = n(A) - n(A \cap B) \qquad \text{and} \qquad n(D) = n(B) - n(A \cap B).$$

Consequently, combining the above equations gives

$$n(A \cup B) = n(A) + n(B) - n(A \cap B).$$

Now let us re-examine infinite sets.

2.3.7 Definition. *An infinite set A is said to be* **countably infinite** *if there exists a one-to-one mapping from A onto the set of positive integers. If no such mapping exists, A is said to be* **uncountably infinite.**

The terms *denumerable* and *nondenumerable* are sometimes used in place of *countable* and *uncountable*, respectively.

First note that the set I of all integers is countably infinite. Observe the following arrangement of the set I and the set P of positive integers:

$$I: \quad 0, 1, -1, 2, -2, 3, -3, \ldots,$$

$$P: \quad 1, 2, 3, 4, 5, 6, 7, \ldots$$

Thus if we define a mapping α from P to I by

$$\alpha(n) = \begin{cases} \dfrac{n}{2} & \text{if } n \text{ is even,} \\ -\dfrac{n-1}{2} & \text{if } n \text{ is odd,} \end{cases}$$

we can see that $I \approx P$. Hence I is countably infinite.

2.3.8 Theorem. *Every infinite subset of a countably infinite set is countably infinite.*

Proof. Let A be a countably infinite set. Label the elements of A such that a_1 is the element which corresponds to 1, a_2 is the element which corresponds to 2, etc. Then $A = \{a_1, a_2, a_3, \ldots\}$. Let B be a subset of A. We must define a one-to-one mapping from the set P of positive integers onto B. Let n_1 be the smallest positive integer such that $a_{n_1} \in B$ and then define $\alpha(1) = a_{n_1}$. Let n_2 be the smallest positive integer greater than n_1 such that $a_{n_2} \in B$ and define $\alpha(2) = a_{n_2}$. Now after $n_1, n_2, \ldots, n_k$ are chosen and $\alpha(1), \ldots, \alpha(k)$ are defined, let n_{k+1} be the smallest positive integer greater than n_k such that $a_{n_{k+1}} \in B$. Then define $\alpha(k + 1) = a_{n_{k+1}}$. By the *Principle of Mathematical Induction* (see 4.1.2), $\alpha(n)$ is defined for all $n \in P$. Moreover, α is one-to-one and onto. Hence $P \approx B$; this proves the theorem.

In the proof of Theorem 2.3.8, we picked n_1 to be the smallest positive integer such that $a_{n_1} \in B$. Do we know whether such an integer exists? We definitely should ask ourselves if it does exist if we are going to use it. The reason such an integer exists is that the set of positive integers with the usual ordering is *well-ordered.* We will discuss this notion in some detail in Section 2.6 and will prove in Section 3.1 that the set of positive integers is well-ordered. The assertion that the set P of positive integers is well-ordered states that every nonempty subset of P has a least (or smallest) element. From our previous experience with the set P, it is easy to convince ourselves that this assertion is believable. Hence we accept it now and wait for the later sections for its proof.

The next theorem proves that the countable union of countably infinite sets is countably infinite.

2.3.9 Theorem. *Let $A_1, A_2, \ldots, A_n, \ldots$ be countably infinite sets and let*

$$A = \bigcup_{n=1}^{\infty} A_n.$$

Then A is countably infinite.

Proof. We must establish a correspondence between A and the set P of positive integers. Each A_n is countably infinite, so we can label the elements of A_n by $a_{n1}, a_{n2}, \ldots, a_{nk}, \ldots$ That is, the kth element of the A_n is a_{nk}. Now consider the following infinite array:

$$\begin{matrix} a_{11} & a_{12} & a_{13} & a_{14} & \cdots \\ a_{21} & a_{22} & a_{23} & a_{24} & \cdots \\ a_{31} & a_{32} & a_{33} & a_{34} & \cdots \\ a_{41} & a_{42} & a_{43} & a_{44} & \cdots \end{matrix}$$

The elements of the nth row are simply A_n. The total array contains all the elements of A. If one follows the arrows, one can arrange the elements of A in succession as follows:

$$\text{(1)} \qquad a_{11}, a_{12}, a_{21}, a_{13}, a_{22}, a_{31}, a_{14}, a_{23}, a_{32}, a_{41}, \cdots$$

This arrangement of the elements of A leads to a one-to-one mapping α from A onto the set of positive integers:

$$\alpha(a_{ij}) = \frac{(i + j - 2)(i + j - 1)}{2} + i.$$

Thus the set $\{a_{ij} \mid i = 1, 2, \ldots, j = 1, 2, \ldots\}$ is countably infinite. If any of the A_n's have elements in common, they will appear more than once in (1). But A is a subset of $\{a_{ij} \mid i = 1, 2, \ldots, j = 1, 2, \ldots\}$; so by Theorem 2.3.8, A is countably infinite. (Note that A is infinite since $A_i \subseteq A$ for all i.)

The next two theorems show that the set of all rational numbers is countably infinite and the set of all real numbers is uncountably infinite.

2.3.10 Theorem. *The set Q of all rational numbers is countably infinite.*

Proof. First we observe that the set S of all pairs (a, b) where $a, b \in I$, $b > 0$, is countably infinite. In fact, if $S_n = \{(a, n) \mid a \in I\}$, then each S_n is countably infinite and

$$S = \bigcup_{n=1}^{\infty} S_n.$$

The set Q of rational numbers is the set of all fractions a/b where $a, b \in I$, $b > 0$. Hence since S is countably infinite, Q is too if we identify the rational number a/b with the ordered pair (a, b).

2.3.11 Theorem. *The set R of all real numbers is uncountably infinite.*

Proof. Suppose that R is countable. Then there exists a one-to-one mapping α from R onto the set P of positive integers. Now we must use a fact we have not proved but which should be quite obvious. Every real number can be written as a decimal (possibly infinite). There are two ways that a number such as 5 can be written as an infinite decimal: 5.00 . . . and 4.999 . . . In cases like this, select one representation to use and use it throughout. Hence each real number can be expressed uniquely in the form $a.a_1a_2a_3 \ldots$, where $a > 1$ and each a_i is 0, 1, 2, . . ., 8, or 9.

Since α is onto P, for each $n \in P$, there exists a real number x such that $\alpha(x) = n$. Let us indicate this x which is mapped onto n by $x = a_n.a_{1n}a_{2n}a_{3n} \ldots$, where a_n is greater than 1 and each a_{in} is 0, 1, 2, . . ., 8, or 9. For instance,

$$\begin{aligned}
&a_1.a_{11}a_{21}a_{31} \ldots a_{n1} \ldots \to 1,\\
&a_2.a_{12}a_{22}a_{32} \ldots a_{n2} \ldots \to 2,\\
&a_3.a_{13}a_{23}a_{33} \ldots a_{n3} \ldots \to 3,\\
&\vdots\\
&a_n.a_{1n}a_{2n}a_{3n} \ldots a_{nn} \ldots \to n.
\end{aligned}$$

Let us now consider the real number $0.a_{11}a_{22}a_{33} \ldots a_{nn} \ldots$ Construct a new real number z as follows: If $a_{ii} = 2$, replace a_{ii} by 1; if $a_{ii} \neq 2$, replace it by 2. Now the real number z with the replaced values for the a_{ii}'s differs from each $a_n.a_{1n}a_{2n} \ldots$ in at least one digit. Hence z is not the image of an element of P, contradicting the fact that α is onto. This contradiction proves that R is uncountable.

2.3.12 Exercises

1. Prove that the set of positive integers and the set of positive integers above 100 have the same cardinality.
2. For any finite sets A and B, prove that $n(A \times B) = n(A)n(B)$.
3. Prove that there are as many positive integers as there are negative integers.
4. Let A and B be finite sets and let α be a mapping from A onto B. Prove that $n(A) \geq n(B)$. When is $n(A) = n(B)$?
5. Let $A = \{\text{real numbers } x \mid 0 < x < 1\}$ and let D denote the set of all positive real numbers. Use the mapping $\alpha: A \to D$ given by $\alpha(x) = 1 + 1/(x - 1)$ for all $x \in A$ to prove that $A \approx D$. Find another mapping which will work.
6. Let A be as in Exercise 5, let $B = \{\text{real numbers } x \mid 0 < x < 2\}$, and let $C = \{\text{real numbers } x \mid 3 < x < 7\}$. Prove that $A \approx B \approx C$.
7. Let S be a set whose elements are countably infinite sets whose elements are either 0 or 1; e.g., a typical element of S would be

$$\{0, 0, 0, 1, 0, 1, 1, 0, \ldots\}.$$

Let E be a countably infinite subset of S. Label the elements of E by $S_1, S_2, \ldots$; hence for each $i = 1, 2, \ldots, S_1$ is a countably infinite set consisting of 0's and 1's. Find a countably infinite set $T \in S$ such that $T \notin E$. Show that the existence of T proves that S is uncountable.

2.4 EQUIVALENCE RELATIONS

From our previous mathematical experience we realize the importance of the concept of equality. The notion of equivalence relation is a generalization of equality which will play an important role in our study of modern algebra.

Intuitively a *relation* $\sim$ on a set A is a mapping which associates with each ordered pair (a, b) of elements of A a statement $a \sim b$ such that, first of all, $a \sim b$ makes sense, and second, that $a \sim b$ is either true or false, *but never both.* A more rigorous definition of relation is given in 2.6.1. Our interest in relations will be restricted to equivalence relations.

2.4.1 Definition. *A relation $\sim$ on a set A is an* ***equivalence relation*** *if and only if $\sim$ has the following properties:*

1) $a \sim a, \forall a \in A$ *(reflexive)*
2) $a \sim b \Rightarrow b \sim a, \forall a, b \in A$ *(symmetric)*
3) $a \sim b \wedge b \sim c \Rightarrow a \sim c, \forall a, b, c \in A$ *(transitive)*

The symbol $a \sim b$ is read "a is equivalent to b" and if a is not equivalent to b, then we write $a \nsim b$.

The relation $<$ on the integers is not an equivalence relation, since it is not reflexive. The relation $\leq$ on the integers is reflexive and transitive, but not symmetric. Hence it also fails to be an equivalence relation. However, if we consider the set of all triangles in the plane, then "is congruent to" and "is similar to" are both equivalence relations on this set of triangles.

Now we consider an example of an important equivalence relation on the set I of all integers. Define

(1) $a \sim b$ if and only if $a - b$ is a multiple of 5.

Sometimes the notation $a \equiv b \pmod 5$ is used rather than $a \sim b$ for this equivalence relation. We will now prove that (1) defines an equivalence relation on I. Let $a, b, c \in I$. Then $a \sim a$ since $a - a = 0 = 0 \cdot 5$; that is, $a - a$ is a multiple of 5. This proves that $\sim$ is reflexive. Now suppose that $a \sim b$. This means that $a - b = 5n$ for some integer n. Hence

$$b - a = -(a - b) = -(5n) = 5(-n),$$

and so $b - a$ is a multiple of 5. Thus $b \sim a$, which proves that $\sim$ is symmetric. Next we show that $\sim$ is transitive. Suppose $a \sim b$ and $b \sim c$. Then $a - b = 5m$ and $b - c = 5k$ for some integers m and k. Then

$$a - c = (a - b) + (b - c) = 5m + 5k = 5(m + k).$$

Hence $a \sim c$. Therefore $\sim$ is an equivalence relation on I.

2.4.2 Definition. *Let $\sim$ be an equivalence relation on a nonempty set A. For each $a \in A$,*

$$[a] = \{x \in A \mid x \sim a\}$$

*is called the **equivalence class of a** and a is called a **representative** of the equivalence class $[a]$.*

Thus $[a]$ is the set of elements $x \in A$ which are equivalent to a. Note that since $a \sim a$ by the reflexive property, $a \in [a]$ for all $a \in A$. Hence $[a]$ is never the empty set.

Let us return now to the equivalence relation on I defined above, that is, $a \sim b$ if and only if $a - b$ is a multiple of 5. What are the equivalence classes associated with this equivalence relation? In particular, what integers belong to $[2]$? If $n \in [2]$, then by definition of equivalence class, $n \sim 2$. Hence $n - 2$ is a multiple of 5. Thus n is an integer of the form $n = 2 + 5k$, where k can be any integer. Thus

$$[2] = \{\ldots, -13, -8, -3, 2, 7, 12, 17, \ldots\}.$$

By a similar procedure we see that

$$\begin{aligned}
[0] &= \{\ldots, -15, -10, -5, 0, 5, 10, 15, \ldots\},\\
[1] &= \{\ldots, -14, -9, -4, 1, 6, 11, 16, \ldots\},\\
[3] &= \{\ldots, -12, -7, -2, 3, 8, 13, 18, \ldots\},\\
[4] &= \{\ldots, -11, -6, -1, 4, 9, 14, 19, \ldots\}.
\end{aligned}$$

By examination of these equivalence classes we see that every integer belongs to one of these five equivalence classes and no two of the equivalence classes have any elements in common. Therefore [0], [1], [2], [3], and [4] form a *partition* of I. We will show that this is a property of equivalence classes in general.

2.4.3 Theorem. *Let $\sim$ be an equivalence relation on a set A. Then the following are true for $a, b \in A$:*

1) If $a \in [b]$, then $[a] = [b]$.
2) $[a] = [b]$ if and only if $a \sim b$.
3) If $[a] \cap [b]$ is not empty, then $a \sim b$.
4) $[a] \cap [b]$ is empty if and only if $a \nsim b$.

Proof. We will prove (1) and leave the rest for exercises. To prove (1), we must show that the two sets $[a]$ and $[b]$ are equal, i.e., that $[a] \subseteq [b]$ and $[b] \subseteq [a]$.

Assume that $a \in [b]$. Then by definition of equivalence classes, $a \sim b$. Now let $x \in [a]$. Then $x \sim a$. Since $x \sim a$ and $a \sim b$, by the transitive property of equivalence relations, $x \sim b$. Hence $x \in [b]$. Consequently, we have shown that every element x belonging to $[a]$ also belongs to $[b]$. Hence $[a] \subseteq [b]$.

Now let $y \in [b]$. Then $y \sim b$. Since $a \sim b$, by the symmetric property, $b \sim a$. Since $y \sim b$ and $b \sim a$, the transitive property implies that $y \sim a$. Hence $y \in [a]$. Therefore $[b] \subseteq [a]$. Finally, since $[a] \subseteq [b]$ and $[b] \subseteq [a]$, we have $[a] = [b]$ as asserted.

As a result of Theorem 2.4.3, we get the following important theorem.

2.4.4 Theorem. *Let $\sim$ be an equivalence relation on a set A. Then the set of equivalence classes of $\sim$ form a partition of A.*

Proof. First, for every element $a \in A$, $a \sim a$ by the reflexive property and so $a \in [a]$. Thus every element of A belongs to at least one equivalence class. Parts (2) and (3) of Theorem 2.4.3 show that no element can belong to two different equivalence classes. Therefore the set of equivalence classes forms a partition of A.

2.4.5 Exercises

1. Let T be the set of all triangles in the plane. Verify that "is similar to" is an equivalence relation on T.
2. Let I be the set of integers and define $\sim$ on I by $m \sim n$ if and only if $m - n$ is a multiple of 4 for all $m, n \in I$. Prove that this is an equivalence relation on I. What are the equivalence classes associated with this equivalence relation? Prove that $[3] = [11]$.
3. Let I be the set of integers and define $\sim$ on I by $a \sim b$ if and only if a is a multiple of b (i.e., $a \sim b$ if and only if $a = bm$ for some $m \in I$) for all $a, b \in I$. Prove or disprove that this is an equivalence relation.
4. Prove parts (2), (3), and (4) of Theorem 2.4.3.
5. Let $\sim$ be an equivalence relation on a set A. If $a, b, c, d \in A$ and $c \in [a]$, $d \in [b]$, and $[a] \neq [b]$, prove that $c \nsim d$.
6. One might ask if it is necessary to have all three conditions in the definition of equivalence relation. By symmetry $a \sim b \Rightarrow b \sim a$ and by transitivity $a \sim b$ and $b \sim a \Rightarrow a \sim a$. Hence it seems that symmetry and transitivity imply reflexivity. So why do we need to require that equivalence relations be reflexive?
7. We defined what is meant by equivalent sets in 2.3.1. Is this really an equivalence relation?
8. Define a relation $\sim$ on the set R of real numbers by

   $$a \sim b \Leftrightarrow a > b \text{ and } a < b.$$

   Is this an equivalence relation?

2.5 BINARY OPERATIONS

The set I of integers is important to us since we have operations—namely, addition, subtraction, and multiplication—which can be used to combine integers to get other integers. Such operations are called *binary operations on I.* Some general sets have binary operations defined on them which allow us to combine two elements of the set to get another element of the set.

2.5.1 Definition. *Let A be a nonempty set. A* ***binary operation*** $\cdot$ ***on*** A *is a correspondence which associates to each ordered pair (a, b) of elements of A a unique element $a \cdot b$ of A.*

An alternative definition of binary operation is this: It is a mapping from the cross product $A \times A$ to A, where the image of (a, b) is $a \cdot b$. There are three words in Definition 2.5.1 which merit extra emphasis. The first is "each." If $\cdot$ is to be a binary operation on a set A, it must define $a \cdot b$ for every pair of elements a, b of A; i.e., there must not be any elements such that we cannot combine them. The next important word is "unique." For each pair of elements a, b of A, there must be only one "answer" $a \cdot b$ when we combine them. In particular, this says that if $a_1, a_2, b_1, b_2 \in A$ and if $a_1 = a_2$ and $b_1 = b_2$, then $a_1 \cdot b_1 = a_2 \cdot b_2$. When the operation in question has this "uniqueness," the operation is said to be *well-defined.* The last word to be stressed is "of." For $\cdot$ to be a binary operation on A, $a \cdot b$ must be an element of A. Hence for every $a, b \in A$, $a \cdot b \in A$. This property is usually called the *closure property.* Division is not a binary operation on the set of integers since it is not *closed*; for example, 3 divided by 4 is not an integer.

A binary operation on a set may have additional properties. There are two in particular that are of interest in modern algebra.

2.5.2 Definition. *Let $\cdot$ be a binary operation on a set A:*

1) $\cdot$ *is* ***commutative*** *if $a \cdot b = b \cdot a$ for all $a, b \in A$.*

2) $\cdot$ *is* ***associative*** *if $(a \cdot b) \cdot c = a \cdot (b \cdot c)$ for all $a, b, c \in A$.*

The binary operations addition $(+)$ and multiplication $(\cdot)$ on the integers are commutative and associative since $a + b = b + a$, $a \cdot b = b \cdot a$, $(a \cdot b) \cdot c = a \cdot (b \cdot c)$, and $(a + b) + c = a + (b + c)$ for all integers a, b, and c. In addition, there is a relationship between $+$ and $\cdot$, namely, $a \cdot (b + c) = (a \cdot b) + (a \cdot c)$ and $(a + b) \cdot c = (a \cdot c) + (b \cdot c)$ for all $a, b, c \in I$. This relationship is extended to sets in general by the following definition.

2.5.3 Definition. *Let $\cdot$ and $*$ be binary operations on a set A:*

1) $*$ *is* ***left distributive over*** $\cdot$ ***if*** $a * (b \cdot c) = (a * b) \cdot (a * c)$ *for all* $a, b, c \in A$.

2) $*$ *is* ***right distributive over*** $\cdot$ ***if*** $(a \cdot b) * c = (a * c) \cdot (b * c)$ *for all* $a, b, c \in A$.

The exercises at the end of the section provide practice in determining whether or not certain binary operations on a set satisfy the properties defined in the foregoing material.

Next we turn to specific elements in a set which has a binary operation defined on it.

2.5.4 Definition. *Let S be a set which has a binary operation $\cdot$ defined on it. An element $e \in S$ is called an* ***identity element with respect to*** $\cdot$ *if $a \cdot e = e \cdot a = a$ for all $a \in S$.*

In general, a set may not have an identity element with respect to a certain binary operation. For instance, the integer zero is the identity element of the set I of integers with respect to the binary operation addition, and the integer one is the identity element of the set I of integers with respect to the binary operation multiplication. On the other hand, the set E of even integers has no identity element with respect to the binary operation multiplication.

If one knows that a set S with binary operation $\cdot$ has an identity element, a natural question to ask is "Does S have more than one identity element?" The answer is "no!" as the following theorem shows.

2.5.5 Theorem. *Let S be a set on which is defined a binary operation $\cdot$. If S has an identity element with respect to $\cdot$, then this identity is unique.*

Before we give the proof of this theorem, let us examine exactly what it is that must be proved. Since we know by hypothesis that S has at least one identity element with respect to $\cdot$, we must prove that it has no more than one. We will assume that S has more than one identity element with respect to $\cdot$ and deduce that they are identical. This method of proving uniqueness is "standard operating procedure."

Proof. Suppose e and f are both identity elements of S with respect to $\cdot$. Then by definition,

$$e \cdot a = a \cdot e = a \tag{1}$$

for all $a \in S$, and

$$f \cdot a = a \cdot f = a \tag{2}$$

for all $a \in S$. In particular (1) must hold true for $a = f$. Thus we have

$$e \cdot f = f \cdot e = f. \tag{3}$$

Similarly (2) must hold true for $a = e$, so

(4) $$f \cdot e = e \cdot f = e.$$

Comparing (3) and (4) one concludes that $e = f$.

2.5.6 Definition. *Let S be a set with binary operation $\cdot$ defined on it. Suppose that S has an identity element e with respect to $\cdot$. Let a be an element of S. An element b of S is said to be an* ***inverse of a with respect to*** $\cdot$ *if $b \cdot a = a \cdot b = e$.*

If one considers the set I of integers, then for $a \in I$, $-a$ is the inverse of a with respect to addition. However, 1 and -1 are the only elements of I which have inverses with respect to multiplication. Also note the difference in quantification in Definitions 2.5.4 and 2.5.6. For $e \in S$ to be an identity element with respect to $\cdot$, $a \cdot e = e \cdot a = a$ must be true for *all* $a \in S$. But b is the inverse of an element $a \in S$ with respect to $\cdot$ if $b \cdot a = a \cdot b = e$. In other words, e must "work" for all $a \in S$, whereas b "works" only for this one a.

2.5.7 Theorem. *Let S be a set with associative binary operation $\cdot$ defined on it. Suppose that S has an identity element e with respect to $\cdot$. Let $a \in S$ and assume that a has an inverse with respect to $\cdot$. Then this inverse is unique.*

We follow a scheme similar to the proof of Theorem 2.5.5.
Proof. Assume that b and c are inverses of a with respect to $\cdot$. Then

(1) $$a \cdot b = b \cdot a = e$$

and

(2) $$a \cdot c = c \cdot a = e.$$

Then
$$\begin{aligned} b &= b \cdot e && (2.5.4) \\ &= b \cdot (a \cdot c) && (2), \\ &= (b \cdot a) \cdot c && \text{(Associativity)}, \\ &= e \cdot c && (1), \\ &= c && (2.5.4). \end{aligned}$$

Therefore $$b = c.$$

2.5.8 Exercises

1. Prove or disprove that the following operations are binary operations on the indicated sets.
   a) The usual addition of integers on the set E of even integers.

b) The usual addition of integers on the set O of odd integers.

c) Let I be the set of integers. For all $a, b \in I$, define a ⓜ b to be any integer larger than both a and b.

d) Let I be the set of nonzero integers. For all $a, b \in I$, define $a \,\$\, b$ to be the least common multiple of a and b.

e) Let S be the set of all subsets of a given nonempty set A. For $B, C \in S$, define $B \cdot C = B \cap C$.

f) Let S be the set of all subsets of a given nonempty set A. For $B, C \in S$, define

$$B \cdot C = \{x \mid x \in B \vee x \in C, \text{ but } x \notin B \cap C\}.$$

2. Examine the following operations on the indicated sets to determine whether or not they are well-defined.

a) Let Q denote the set of rational numbers a/b, where a and b are integers, $b \neq 0$. Two rational numbers a/b and c/d are said to be *equal* if $ad = bc$. Define $\oplus$ on Q by

$$\frac{a}{b} \oplus \frac{c}{d} = \frac{a + c}{b + d}.$$

b) With Q as in (a), define $\otimes$ on Q by

$$\frac{a}{b} \otimes \frac{c}{d} = \frac{ac}{bd}.$$

c) With Q as in (a), define $\circledast$ on Q by

$$\frac{a}{b} \circledast \frac{c}{d} = \frac{ad + bc}{bd}.$$

d) Let $\sim$ be the equivalence relation on the set I of integers given by $a \sim b$ if and only if $a - b$ is a multiple of 5. Let J be the set of equivalence classes. Define $\oplus$ on J by

$$[m] \oplus [n] = [m + n].$$

e) With J as in (d), define $\odot$ by

$$[m] \odot [n] = [mn].$$

f) With J as in (d), define $*$ by

$$[m] * [n] = [m + n + 1].$$

3. In each of the following, a set is given and a binary operation is defined on it. Prove or disprove that the binary operation is commutative and/or associative. Let I be the set of integers throughout this exercise.

a) I; $a \cdot b = a - b$ for all $a, b \in I$

b) I; $a \# b = a + b + 1$ for all $a, b \in I$

c) I; $a \oplus b = a + b - ab$ for all $a, b \in I$

d) $T = \{a + b\sqrt{2} \mid a, b \in I\}$; usual multiplication of real numbers

e) E = set of even integers; $a \cdot b = 2ab$ for all $a, b \in E$

f) S = set of subsets of A; $B \cdot C = B \cap C$ for all $B, C \in S$

g) S = set of subsets of A;

$$B \cdot C = \{x \in A \mid x \in B \vee x \in C, \quad \text{but } x \notin B \cap C\}$$

4. Let I be the set of nonzero integers. Prove or disprove that multiplication is left distributive over each of the following:

   a) $a \cdot b = a - b$ for all $a, b \in I$
   b) $a \# b = a + b + 1$ for all $a, b \in I$
   c) $a \oplus b = a + b - ab$ for all $a, b \in I$
   d) $a \$ b$ = least common multiple of a and b for all $a, b \in I$

5. Let S be a given set with a binary operation $\cdot$ defined on it. Let A and B be subsets of S such that $\cdot$ is commutative and associative on A and on B.
   a) Prove that $\cdot$ is commutative on $A \cap B$.
   b) Prove that $\cdot$ is associative on $A \cap B$.

6. Which, if any, of the sets in Exercise 3 have identity elements with respect to the binary operation defined there?

7. Let H be a set with operation $+$ defined on H which satisfies the following axioms:

   i) For all $x, y \in H$, $x + y \in H$.
   ii) For all $x, y, z \in H$, $(x + y) + z = (x + z) + y$.
   iii) There exists $w \in H$ such that $w + x = x$ for all $x \in H$.

   Prove all the theorems you can about H. Give an example of a system of which H is an abstraction.

2.6 ORDERINGS

All students realize what is meant by the expression "a is less than b" where a and b are certain real numbers. This is an example of an *order relation.* In general, a given set may not have an order relation defined on it; however, many do. In this section we define and prove some elementary properties of sets having order relations defined on them.

First of all, let us give a more precise definition of a relation on a set.

2.6.1 Definition. *Let A be a set. A* ***relation on*** A *is a subset R of $A \times A$; i.e., a relation on a set A is a certain subset R of all ordered pairs of elements of A. If $(a, b) \in R$, where $a, b \in A$, we say that a is related to b and write aRb or $a \sim b$.*

2.6.2 Definition. *Let A be a set. A relation R on A is a* ***partial order*** *(or* ***partial ordering****)* ***on*** A *if and only if R is transitive on A, i.e., if aRb and bRc imply aRc for all $a, b, c \in A$.*

We often use the symbol $<$ in place of R to designate that two elements of a set are related; i.e., we write $a < b$ if aRb. In this case we say "a is less than b," "a precedes b," or "b is greater than a." However, the particular partial ordering which is denoted by $<$ may not be the one with which we are accustomed.

The relation "is older than" is a partial ordering on the set of all people. Since polygamy still exists in some cultures, "is married to" is not a partial ordering on the set of all people. If we set $a < b$ when $b - a$ is positive, then $<$ is a partial ordering on the set of all integers. Let A be a given nonempty set and let A^* be the set of all subsets of A. For elements $M, N \in A^*$, set $M < N$ if $M \subseteq N$. This is a partial order on A^*.

2.6.3 Definition. *Let $<$ be a partial ordering on a set A. Then $<$ is a* ***linear ordering (total ordering, ordering)*** *if the following hold for all $x, y \in A$:*

1) If $x < y$ and $y < x$, then $x = y$.
2) If $x \neq y$, then $x < y$ or $y < x$.

Note that linear orderings (or partial orderings, for that matter) need not be reflexive or symmetric. In fact (1) states that a linear ordering is extremely nonsymmetric, for the only time that $x < y$ implies $y < x$ is when $x = y$. Condition (2) is usually called the *trichotomy law*. For a set A, the partial relation defined above on A^* is not a linear ordering since it does not satisfy (2). For instance, if $A = \{a, b, c, d, e\}$, $M = \{a, b, c\}$, and $N = \{c, d, e\}$, then $M, N \in A^*$ but $M \neq N$, $M \nsubseteq N$, and $N \nsubseteq M$. An intuitive way of expressing the fact that a relation on a set A satisfies the trichotomy law is to say that every element of A is related to every other element of A.

For any set A, we use the notation $a \leq b$ for elements $a, b \in A$ to mean $a < b$ or $a = b$.

Certain sets upon which a linear ordering has been defined have an additional property which is rather important. However, before stating it, we must consider the following definition.

2.6.4 Definition. *Let A be a set and let $<$ be a linear ordering on A. An element $x \in A$ is said to be the* ***least element of A*** *if $x \leq b$ for all $b \in A$.*

Note that x must be an element of A in order to be the least element of A.

2.6.5 Definition. *Let A be a set and let $<$ be a linear ordering on A. The set A with the ordering $<$ is said to be* **well-ordered** *if and only if every nonempty subset of A has a least element.*

The set P of positive integers with the usual ordering is well-ordered. Intuitively this statement seems rather obvious. In Chapter 3 a detailed proof of this fact is given. From this one can readily deduce that the set N of nonnegative integers with the usual ordering is well-ordered. The fact that P and N, with their usual orderings, are well-ordered plays an important role in the proofs of many theorems in mathematics. Several proofs in later chapters of this book use this fact; see, for example, 4.1.1 and 4.2.3. On the other hand, the set I of all integers with the usual ordering is not well-ordered, since the set I itself has no least element. What about the set Q_p of positive rationals with the usual ordering? It is easy to see that the subset $\{1/2, 1/3, \ldots, 1/n, \ldots\}$ has no smallest element. Hence Q_p is not well-ordered when Q_p has its usual ordering.

When one speaks of an ordered set, it is necessary to specify the order relation and the set, because a given set may have several order relations defined upon it. In particular, in Definition 2.6.5 when "well-ordered" is defined it is in terms of a set A *and* the linear order $<$. It quite often happens that a set may be well-ordered with one ordering and not well-ordered in another. In fact, if the *Axiom of Choice** is assumed, then it can be proved that every nonempty set has an ordering defined on it so that the set is well-ordered in that ordering. Thus there exist orderings on I and on Q_p such that I and Q_p are well-ordered with these orderings. Can you find these orderings?

It is easy to see that the set I of all integers with the usual ordering is not well-ordered, since the set I itself does not have a least element. One might guess that the set Q_p of positive rational numbers with the usual ordering is well-ordered. However, this guess is incorrect; for consider the subset $\{1/2, 1/3, \ldots, 1/n, \ldots\}$ of Q_p. This set has no smallest element.

The fact that N and P are well-ordered is the basis for many important results in mathematics. We will use this fact quite often in the proof of theorems later in this book.

---

* The Axiom of Choice simply states that given a collection of nonempty sets, it is possible to pick one element from each of the sets. The Axiom of Choice is, in fact, equivalent to the well-ordering principle which states that every set can be well-ordered with some ordering. For further discussion on this matter, see the references at the end of the chapter.

2.6.6 Exercises

1. Let U denote the set of people in the United States. Which of the following relations on U are partial orderings? Which are linear orderings?

   a) Is married to
   b) Is younger than
   c) Is taller than
   d) Is the same sex as
   e) Is a friend of
   f) Is a parent of
   g) Is a descendant of

2. Let A and B be sets upon which there are defined partial orderings. Denote both orderings by $<$. Let α be a mapping from A to B. Then α is said to *preserve* the partial ordering if, for any two elements $x, y \in A$, $x < y$ implies $\alpha(x) < \alpha(y)$. Similarly, α is said to *reverse* the partial ordering if, for any two elements $x, y \in A$, $x < y$ implies $\alpha(x) > \alpha(y)$. Consider the following mappings from the set I of all integers into itself. Determine which mappings preserve the usual partial ordering on I and which reverse it.

   a) $\alpha(i) = 3i - 7$ for all $i \in I$
   b) $\alpha(i) = 1 - 3i$ for all $i \in I$
   c) $\alpha(i) = i^2$ for all $i \in I$
   d) $\alpha(i) = i^3$ for all $i \in I$
   e) $\alpha(i) = -i^2$ for all $i \in I$

3. Consider the set I of all integers. Define a partial ordering on I by the following: for $x, y \in I$, $x < y$ if x divides into y with remainder zero; i.e., $x < y$ if x is a factor of y. For example, $3 < 12$ but $3 \nless 4$. Determine whether or not this is a linear ordering on I.

4. Define an ordering on the set Q_p of positive rationals so that Q_p with this new ordering is well-ordered.

ADDITIONAL REFERENCES FOR CHAPTER 2

Fraenkel, A. A., *Set Theory and Logic*, Reading, Mass.: Addison-Wesley (1966).

Halmos, Paul R., *Naive Set Theory*, Princeton: Van Nostrand (1960).

Kamke, A., *Theory of Sets*, New York: Dover Publications (1950).

Kelley, John L., *General Topology*, Princeton: Van Nostrand (1955). (See Chapter 0 for Axiom of Choice.)

Kemeny, John G., Hazleton Mirkil, J. Laurie Snell, and Gerald L. Thompson, *Finite Mathematical Structures*, Englewood Cliffs, N.J.: Prentice-Hall (1958).

May, Kenneth O., *Elements of Modern Mathematics*, Reading, Mass.: Addison-Wesley (1959).

McCoy, Neal H., *Introduction to Modern Algebra*, Boston: Allyn and Bacon (1960).

Ohmer, Merlin M., Clayton V. Aucoin, and Marion J. Cortez, *Elementary Contemporary Mathematics*, New York: Blaisdell (1964).

Suppes, Patrick, *Axiomatic Set Theory*, Princeton: Van Nostrand (1960).

Tarski, A., *Introduction to Logic*, 2nd rev. ed., New York: Oxford University Press (1946).

3 our number systems

Much of our mathematics depends upon the number systems that we use. In fact, our elementary courses in arithmetic are devoted to a study of the properties of these number systems. In particular, the system of real numbers is the basis of all the mathematics which governs the world in which we live. We have accepted the real numbers and their properties simply because we have been told that they exist and that the properties are true. However, it is important that we now prove through logical inference that the system of real numbers exists and that the properties of the real numbers which have been assumed by us are in fact valid.

The construction of the real numbers involves a series of constructions, each of which is important in itself. First one obtains the set $\mathscr{N}$ of natural numbers, then the set $\mathscr{I}$ of integers. The set $\mathscr{Q}$ of rational numbers can then be obtained from $\mathscr{I}$, and finally the system $\mathscr{R}$ of real numbers is obtained from $\mathscr{Q}$. For the sake of completeness, the construction of the set $\mathscr{C}$ of complex numbers from $\mathscr{R}$ is sketched also.

At each stage in this development we first construct a set. Then we must define an addition and multiplication on that set and develop the properties of the addition and multiplication as theorems. Next we define an ordering on the set and determine its properties. The proofs of many of these theorems relating to the properties of addition, multiplication, and the ordering are left as exercises. When we speak of a *number system*, we will mean a set, say $\mathscr{S}$, together with two binary operations, addition and multiplication, denoted by $+$ and $\cdot$, respectively, and an order relation $<$, all defined on $\mathscr{S}$. We use the following notation $(\mathscr{S}, +, \cdot, <)$ to indicate the number system described above.

Even though the constructions and operations involved in constructing the number systems may seem quite removed from our usual notion of these number systems, the end product is the same. That is, the systems we will develop here from an axiomatic basis are precisely the systems we use in everyday arithmetic. It is an important duty of the student to investigate the axiomatic development in sufficient detail so that he can understand the relationships between the axiomatic concepts and the number systems with which he is familiar. In particular, the student should always carry out sample calculations to convince himself that the operations are the same as the ones in general use.

3.1 THE NATURAL NUMBERS

We want to construct the system of natural numbers, assuming as little as possible. We will start with a system of axioms and, by making the proper definitions, we will obtain a system of numbers which we call the system of *natural numbers* $\{1, 2, \ldots\}$. One standard set of axioms which is used as a basis of the system of natural numbers is called *Peano's Postulates*, since they are attributed to the Italian mathematician Giuseppe Peano (1858–1932). We assume that there exists a set $\mathcal{N}$, called the set of *natural numbers*, which satisfies the following axioms.

3.1.1 Peano's Postulates

1) $1 \in \mathcal{N}$.

2) There exists a one-to-one mapping from $\mathcal{N}$ onto the set of elements of $\mathcal{N}$ different from 1. *The image of $n \in \mathcal{N}$ is denoted by n^* and is called the* ***successor*** *of n.*

3) For any set $M \subseteq \mathcal{N}$ if $1 \in M$ and if $n \in M$ implies $n^* \in M$, then $M = \mathcal{N}$. (This is known as the ***induction axiom.****)*

Intuitively n^* may be thought of as $n + 1$; in fact, we can define n^* to be $n + 1$. In other words, n^* is the natural number following n.

We now turn our attention to defining an addition on $\mathcal{N}$. If our axiomatic system is to have the same properties as the system we usually refer to as the natural numbers, this addition must be associative. In particular, we must at least have $m + (n + 1) = (m + n) + 1$ for all $m, n \in \mathcal{N}$. In the language of successors, we must have $m + n^* = (m + n)^*$. The following theorem shows that this condition is sufficient to define an addition on $\mathcal{N}$.

3.1.2 Theorem. *There is exactly one binary operation + on $\mathcal{N}$ such that $n^* = n + 1$ and $m + n^* = (m + n)^*$ for all $m, n \in \mathcal{N}$.*

Proof. Suppose there are two additions, denoted by $+$ and $\oplus$, on $\mathcal{N}$ such that $n^* = n + 1$, $n^* = n \oplus 1$, $m + n^* = (m + n)^*$, $m \oplus n^* = (m \oplus n)^*$ for all $m, n \in \mathcal{N}$. We must show that $m + n = m \oplus n$ for all $m, n \in \mathcal{N}$; that will prove that $+$ and $\oplus$ are the same binary operation. Let $M = \{m \in \mathcal{N} \mid m + n = m \oplus n \text{ for all } n \in \mathcal{N}\}$. We want to prove that $M = \mathcal{N}$. We use the induction axiom for this purpose. First note that $1 \in M$ since $m + 1 = m^* = m \oplus 1$ by hypothesis. Suppose $n \in M$; i.e., suppose $m + n = m \oplus n$. Then

$$m + n^* = (m + n)^* = (m \oplus n)^* = m \oplus n^*.$$

Consequently, $n \in M$ implies $n^* \in M$. Therefore, by the induction axiom, $M = \mathcal{N}$. This proves the theorem.

This theorem leads us to the following definition.

3.1.3 Definition. *The* **addition** $+$ *on* $\mathcal{N}$ *is the unique binary operation such that* $n^* = n + 1$ *and* $m + n^* = (m + n)^*$ *for all* $m, n \in \mathcal{N}$.

What about multiplication now? Surely we will want $n \cdot 1 = n$ for all $n \in \mathcal{N}$. Moreover, if multiplication is to be distributive over addition, we must have $m \cdot (n + 1) = (m \cdot n) + (m \cdot 1)$; i.e., we must have $m \cdot n^* = m \cdot n + m$. It is left as an exercise to show that there is exactly one binary operation $\cdot$ on $\mathcal{N}$ which satisfies these conditions. Having established that fact, we are in a position to make the following definition.

3.1.4 Definition. *The* **multiplication** $\cdot$ *on* $\mathcal{N}$ *is the unique binary operation such that* $n \cdot 1 = n$ *and* $m \cdot n^* = m \cdot n + m$ *for all* $m, n \in \mathcal{N}$.

The natural number $m \cdot n$ will often be denoted by mn. Also we assume that $\cdot$ takes precedence over $+$; so $mn + p$ means $(mn) + p$.

Now that we have defined $+$ and $\cdot$ on $\mathcal{N}$, let us list and verify their properties.

3.1.5 Theorem. *The following are true for all natural numbers* m, n, *and* p:

1) $(m + n) + p = m + (n + p)$ *(Associative Law of Addition)*
2) $m + n = n + m$ *(Commutative Law of Addition)*
3) $(mn)p = m(np)$ *(Associative Law of Multiplication)*
4) $mn = nm$ *(Commutative Law of Multiplication)*
5) $m(n + p) = mn + mp$ *(Distributive Law)*
6) If $m + p = n + p$, *then* $m = n$ *(Cancellation Law of Addition)*
7) If $mp = np$, *then* $m = n$ *(Cancellation Law of Multiplication)*

We will prove (1) and (6), simply sketch a proof of (2), and leave the rest as exercises.

Proof of (1). Let

$$M = \{p \in \mathcal{N} \mid (n + m) + p = n + (m + p) \text{ for all } m, n \in \mathcal{N}\}.$$

We will use axiom (3) to prove that $M = \mathcal{N}$, thus proving the associativity of $+$ on $\mathcal{N}$. Note that $1 \in M$, since $(n + m) + 1 = (n + m)^* = n + m^* = n + (m + 1)$. Now assume that $p \in \mathcal{N}$; i.e., assume that $(n + m) + p = n + (m + p)$. Then

$$\begin{aligned}(n + m) + (p + 1) &= (n + m) + p^* \\ &= ((n + m) + p)^* = (n + (m + p))^* \\ &= n + (m + p)^* = n + (m + p^*) \\ &= n + (m + (p + 1)).\end{aligned}$$

Hence $p + 1 \in M$. Axiom (3) then implies that $M = \mathcal{N}$.

Sketch of proof of (2). Let

$$M = \{n \in \mathcal{N} \mid n + m = m + n \text{ for all } m \in \mathcal{N}\}.$$

We wish to show that $1 \in M$. However, to show that $1 \in M$ requires a separate induction argument based on axiom (3). When that is completed, the proof of the commutativity of $+$ is similar to the above proof.

Proof of (6). First observe that since the mapping of axiom (2) is one-to-one, $n^* = m^*$ implies $n = m$. Thus if we set

$$M = \{p \in \mathcal{N} \mid n + p = m + p \text{ implies } n = m \text{ for all } n, m \in \mathcal{N}\},$$

then $1 \in M$. Suppose $p \in M$; i.e., suppose $n + p = m + p$ implies $n = m$. Let $n + p^* = m + p^*$. Then $(n + p)^* = (m + p)^*$ by the definition of addition. Since the mapping is one-to-one, $n + p = m + p$. Our assumption now implies that $n = m$. Hence $n + p^* = m + p^*$ implies $n = m$ and therefore $p^* \in M$. By the induction axiom, $M = \mathcal{N}$ which proves (6).

Having defined addition and multiplication and established their basic properties, we turn our attention to ordering on $\mathcal{N}$.

3.1.6 Definition. *If $m, n \in \mathcal{N}$, then $m < n$ if and only if there exists $p \in \mathcal{N}$ such that $m + p = n$.*

We use $m < n$ and $n > m$ interchangeably. Also $m \leq n$ means that $m < n$ or $m = n$. This definition gives an *ordering* on $\mathcal{N}$.

3.1.7 Theorem. *Let $m, n, p \in \mathcal{N}$. Then the following hold.*

1) $<$ *is transitive, but not reflexive or symmetric.*
2) $n < n^*$ *and* $1 \leq n$.
3) Exactly one of the following is true: $m = n$, $m < n$, $n < m$. *(This is the trichotomy law.)*
4) $n < m$ *if and only if* $n + p < m + p$.
5) $n < m$ *if and only if* $np < mp$.
6) If $p \leq m \leq p^*$, *then* $p = m$ *or* $p^* = m$.

Here we prove only (4), leaving the rest as exercises.

Proof of (4). Observe that this is an "if and only if" statement. Suppose first that $n < m$. Then there exists $q \in \mathcal{N}$ such that $n + q = m$. Then since $+$ is a binary operation on $\mathcal{N}$, $(n + q) + p = m + p$ for all $p \in \mathcal{N}$. Using the associativity and commutativity of $+$, we see that $(n + p) + q = m + p$. Hence $n + p < m + p$.

Conversely, now assume that $n + p < m + p$. Then there is a $q \in \mathcal{N}$ such that $(n + p) + q = m + p$. As above, this means $(n + q) + p = m + p$, and if we apply the Cancellation Law of Addition (3.1.5(6)), we see that $n + q = m$. Hence $n < m$.

The system of natural numbers has an important property which is used extensively throughout mathematics. This property is that $\mathcal{N}$ is *well-ordered* with the ordering of 3.1.6. First some definitions are needed.

3.1.8 Definition.

1) Let $M \subseteq \mathcal{N}$. *Then* $a \in M$ *is a* **least element** *of* M *if* $a \leq b$ *for all* $b \in M$.
2) A set X *together with an ordering is* **well-ordered** *if every nonempty subset of* X *has a least element.*

3.1.9 Theorem. $\mathcal{N}$ *with the ordering we have defined here is well-ordered.*

Proof. Let M be a nonempty subset of $\mathcal{N}$. We must show that M has a least element. Let $K = \{k \mid k \in \mathcal{N}, k \leq m \text{ for all } m \in M\}$. Since M is nonempty, there exists $m \in M$. Then $m^* \notin K$, since $m^* > m$ by Theorem 3.1.7(2). Hence $K \neq \mathcal{N}$. But $1 \in K$ since $1 \leq n$ for all $n \in \mathcal{N}$, also by Theorem 3.1.7(2). Hence there exists $k \in K$ such that $k^* \notin K$; for if such a k did not exist, then $K = \mathcal{N}$, which is not true. Since $k^* \notin K$, there must exist $m \in M$ such that $k^* > m$. But since

$k \in K$ we must have $k \leq m$. Hence from Theorem 3.1.7(6), we see that $k = m$. Hence $k \in M$ and $k \leq m$ for all $m \in M$. Therefore k is the least element of M.

Thus we have the natural numbers and some of their properties. Henceforth when we mention the *system $\mathscr{N}$ of natural numbers* it will be understood that we mean the system consisting of $\mathscr{N}$, $+$, $\cdot$, and $<$ as we have defined them above.

3.1.10 Exercises

1. a) Prove that $1 + n = n + 1$ for all $n \in \mathscr{N}$.
   b) Use (a) to prove $n + m = m + n$ for all $m, n \in \mathscr{N}$.
   c) Prove that $\cdot$ is left- and right-distributive over $+$.
   d) Prove that $\cdot$ is associative and commutative. [*Hint:* Use (c).]
   e) Prove the Cancellation Law of Multiplication.
   f) Prove that $n^* \neq n$ for each $n \in \mathscr{N}$.
   g) Prove that $n \neq m + n$ for all $m, n \in \mathscr{N}$.
2. Prove all parts of Theorem 3.1.7, except (4).
3. Prove that there is one and only one multiplication on $\mathscr{N}$ which satisfies $n \cdot 1 = n$ and $n \cdot m^* = (n \cdot m) + n$ for all $m, n \in \mathscr{N}$.

3.2 THE INTEGERS

Intuitively, the way one obtains the set $\mathscr{I}$ of integers from $\mathscr{N}$ is to consider the difference of two natural numbers. Each integer can be expressed as the difference of two natural numbers and, conversely, the difference of two natural numbers is an integer. However, this representation of the integers is not unique; for $-3 = 2 - 5 = 7 - 10$ and $4 = 9 - 5 = 114 - 110$. Hence we must consider a certain integer r to be represented by all pairs of natural numbers whose difference is r; hence we must consider the pairs (m_1, m_2) and (n_1, n_2) of natural numbers to be equivalent if $m_1 - m_2 = n_1 - n_2$. However, in truth we cannot subtract natural numbers. We avoid this apparent dilemma by saying that (m_1, m_2) and (n_1, n_2) are equivalent if $m_1 + n_2 = m_2 + n_1$.

3.2.1 Definition. *Let $\mathscr{K} = \{(m, n) \mid m, n \in \mathscr{N}\}$. Define $\sim$ on $\mathscr{K}$ by $(m_1, m_2) \sim (n_1, n_2)$ if and only if $m_1 + n_2 = m_2 + n_1$ for all (m_1, m_2) and $(n_1, n_2) \in \mathscr{K}$.*

3.2.2 Theorem. *Then $\sim$ is an equivalence relation on $\mathscr{K}$.*

Proof. Exercise.

3.2.3 Definition. *Denote the equivalence class of* $(m, n) \in \mathscr{K}$ *by* $[m, n]$. *Let* $\mathscr{I}$ *be the set of all equivalence classes, that is,* $\mathscr{I} = \{[m, n] \mid m, n \in \mathscr{N}\}$. *Then* $\mathscr{I}$ *is called the set of* ***integers****. Let* $[m_1, m_2]$ *and* $[n_1, n_2]$ *be elements of* $\mathscr{I}$. *Define* $+$ *and* $\cdot$ *by*

$$[m_1, m_2] + [n_1, m_2] = [m_1 + n_1, m_2 + n_2]$$

and

$$[m_1, m_2] \cdot [n_1, n_2] = [m_1 n_1 + m_2 n_2, m_1 n_2 + n_1 m_2].$$

Before continuing, let us illustrate this addition and multiplication by an example. Let $3 = [12, 9]$ and $-2 = [4, 6]$. Then

$$\begin{aligned} 3 + (-2) &= [12, 9] + [4, 6] \\ &= [12 + 4, 9 + 6] \\ &= [16, 15] \\ &= 1, \end{aligned}$$

and

$$\begin{aligned} 3 \cdot (-2) &= [12, 9] \cdot [4, 6] \\ &= [12 \cdot 4 + 9 \cdot 6, 12 \cdot 6 + 4 \cdot 9] \\ &= [102, 108] \\ &= -6. \end{aligned}$$

Since the operations $+$ and $\cdot$ are defined on equivalence classes, we must verify that they are well-defined. Suppose $[m_1, m_2] = [m_1', m_2']$ and $[n_1, n_2] = [n_1', n_2']$. Then

$$m_1 + m_2' = m_1' + m_2 \qquad \text{and} \qquad n_1 + n_2' = n_1' + n_2.$$

Hence by adding and applying the commutative and associative laws of addition of natural numbers, we see that

$$(m_1 + n_1) + (m_2' + n_2') = (m_1' + n_1') + (m_2 + n_2).$$

Therefore

$$[m_1 + n_1, m_2 + n_2] = [m_1' + n_1', m_2' + n_2'],$$

that is,

$$[m_1, m_2] + [n_1, n_2] = [m_1', m_2'] + [n_1', n_2'],$$

which shows that the addition is well-defined. That multiplication is also well-defined is left as an exercise.

Quite often one denotes an element of $\mathscr{I}$ by a single letter; for instance, one might say $a \in \mathscr{I}$. This simply means that a is an equivalence class of the form $[m, n]$ for some $m, n \in \mathscr{N}$. This notation simplifies the statement of properties of addition and multiplication on integers.

3.2.4 Theorem. *Let $a, b, c \in \mathscr{I}$. Then the following rules hold:*

1) $(a + b) + c = a + (b + c)$ (*Associative Law of Addition*)
2) $a + b = b + a$ (*Commutative Law of Addition*)
3) *There exists an element* $0 \in \mathscr{I}$ *such that for every* $a \in \mathscr{I}$, $a + 0 = a$ (*Existence of Additive Identity*)
4) *For each* $a \in \mathscr{I}$, *there exists* $-a \in \mathscr{I}$ *such that* $a + (-a) = 0$ (*Existence of Additive Inverses*)
5) $(ab)c = a(bc)$ (*Associative Law of Multiplication*)
6) $ab = ba$ (*Commutative Law of Multiplication*)
7) *There exists* $e \in \mathscr{I}$ *such that for all* $a \in \mathscr{I}$ $a \cdot e = a$ (*Existence of Multiplicative Identity*)
8) $a(b + c) = ab + ac$ (*Distributive Law*)

Proofs will be given for (1) and (6), and the remaining parts will be proved as exercises.

Proof of (1). Let $a = [a_1, a_2]$, $b = [b_1, b_2]$, and $c = [c_1, c_2]$ for some natural numbers a_1, a_2, b_1, b_2, c_1, and c_2. Then

$$\begin{aligned}(a + b) + c &= [a_1 + b_1, a_2 + b_2] + [c_1, c_2] \\ &= [(a_1 + b_1) + c_1, (a_2 + b_2) + c_2] \\ &= [a_1 + (b_1 + c_1), a_2 + (b_2 + c_2)]\end{aligned}$$

by the associative law of addition for natural numbers. Hence

$$\begin{aligned}(a + b) + c &= [a_1, a_2] + [b_1 + c_1, b_2 + c_2] \\ &= a + (b + c).\end{aligned}$$

Proof of (6). Let $a = [a_1, a_2]$ and $b = [b_1, b_2]$ for some natural numbers a_1, a_2, b_1, and b_2. Then

$$\begin{aligned}ab &= [a_1, a_2] \cdot [b_1, b_2] \\ &= [a_1b_1 + a_2b_2, a_1b_2 + a_2b_1]\end{aligned}$$

and

$$\begin{aligned}ba &= [b_1, b_2] \cdot [a_1, a_2] \\ &= [b_1 a_1 + b_2a_2, b_1a_2 + b_2a_1].\end{aligned}$$

Hence, using the commutative laws of addition and multiplication of natural numbers, we see that $ab = ba$.

Let us examine what we have accomplished so far. From Peano's Postulates we were given a set $\mathscr{N}$ of natural numbers. We defined an addition and multiplication on this set and, using only the axioms we had stated, we proved theorems listing properties of these binary operations on $\mathscr{N}$. Next we defined an ordering on $\mathscr{N}$ and proved several properties associated with it. The system $(\mathscr{N}, +, \cdot, <)$ based entirely upon axioms and our deductive logic has properties which we naturally associated with our intuitive notion of the natural numbers. This system had several seeming defects—one in particular is that no element of $\mathcal{N}$ has an additive inverse.

Consequently, we turn to a new and "bigger" system. In this present section we have constructed the set $\mathcal{I}$ of integers using only the natural numbers and the properties of their binary operations and order relation. Again we proved theorems listing the properties of addition and multiplication on $\mathcal{I}$. Next we must define an ordering on $\mathcal{I}$. Thinking of the integers $[m_1, m_2]$ and $[n_1, n_2]$ as the differences $m_1 - m_2$ and $n_1 - n_2$, it is intuitively clear that the correct definition of an ordering $<$ on $\mathcal{I}$ must be $[m_1, m_2] < [n_1, n_2]$ if and only if $m_1 - m_2 < n_1 - n_2$. But again, since we cannot subtract natural numbers, we use the following definition.

3.2.5 Definition. *Define $<$ on $\mathcal{I}$ by $[m_1, m_2] < [n_1, n_2]$ if and only if $m_1 + n_2 < m_2 + n_1$ for all $[m_1, m_2]$ and $[n_1, n_2] \in \mathcal{I}$.*

Even though we use the same symbols $+$, $\cdot$, and $<$ for the operations and ordering on both $\mathcal{N}$ and $\mathcal{I}$, there should be no confusion. If the elements involved are natural numbers, then $+$, $\cdot$, and $<$ are those defined on $\mathcal{N}$; on the other hand, when integers are involved, by $+$, $\cdot$, and $<$ we mean the operations and ordering we have just defined.

When we write $m_1 + n_2 < m_2 + n_1$ in Definition 3.2.5, the ordering $<$ is the ordering which we have previously defined on the natural numbers $\mathcal{N}$. In Definition 3.2.5 we have defined $<$ on equivalence classes and consequently we must concern ourselves with the question of whether or not $<$ is well-defined. Suppose that $[m_1, m_2] = [m_1', m_2']$ and $[n_1, n_2] = [n_1', n_2']$. Then it is a rather routine matter to prove that $m_1 + n_2 < m_2 + n_1$ if and only if $m_1' + n_2' < m_2' + n_1'$. This proves that $<$ is indeed well-defined. Next we list some of the properties of $<$ on $\mathcal{I}$, prove some, and leave the proof of others as an exercise. It should be no surprise that the proofs of the properties of the order relation $<$ on $\mathcal{I}$ depend heavily on the properties of the ordering $<$ defined on $\mathcal{N}$.

3.2.6 Theorem. *The following properties hold for all $a, b, c \in \mathcal{I}$:*

1) *Exactly one of the following is true: $a = b$, $a < b$, or $a > b$. (This is the trichotomy law for $\mathcal{I}$.)*
2) *$a < b$ if and only if $a + c < b + c$.*
3) *Let $a > 0$. Then $b < c$ if and only if $ab < ac$.*
4) *Let $a < 0$. Then $b < c$ if and only if $ab > ac$.*

Proof of (2). Let $a = [a_1, a_2]$, $b = [b_1, b_2]$, and $c = [c_1, c_2]$ for natural numbers a_1, a_2, b_1, b_2, c_1, and c_2. Suppose first that $a < b$. Then $a_1 + b_2 < a_2 + b_1$. Hence by Theorem 3.1.7(4),

$$(a_1 + b_2) + (c_1 + c_2) < (a_2 + b_1) + (c_1 + c_2).$$

Thus $(a_1 + c_1) + (b_2 + c_2) < (a_2 + c_2) + (b_1 + c_1)$ and so

$$[a_1 + c_1, a_2 + c_2] < [b_1 + c_1, b_2 + c_2],$$

or in other words, $a + c < b + c$. Now if we suppose that $a + c < b + c$, we have only to apply Theorem 3.1.7(4) again to show that $a < b$.

Hint for proofs of (3) and (4). Let $a = [a_1, a_2]$ for natural numbers a_1, a_2. By 3.2.8, Exercise 3(b), $0 = [n, n]$ for any natural number n. Thus $a > 0$ if and only if $a_1 + n > a_2 + n$. Applying Theorem 3.1.7(4), we see that $a > 0$ if and only if $a_1 > a_2$. Similarly $a < 0$ if and only if $a_2 > a_1$.

In our usual conception of the system of natural numbers and the system of integers, the set of integers contains the set of natural numbers as a subset. But as we have constructed the set of integers, this is not strictly the case. Indeed, each integer is an equivalence class of ordered pairs of natural numbers, as a subset of the set of integers. This process is called the *embedding* of $\mathscr{N}$ in $\mathscr{I}$. We digress now from our study of number systems to discuss embeddings.

3.2.7 Definition. *Let $(\mathscr{S}, +, \cdot, <)$ and $(\mathscr{T}, +, \cdot, <)$ be two number systems consisting of sets $\mathscr{S}$ and $\mathscr{T}$, respectively, with addition $+$, multiplication $\cdot$, and order relation $<$. Then we say that $(\mathscr{S}, +, \cdot, <)$ is* ***embedded*** *in $(\mathscr{T}, +, \cdot, <)$ if and only if there exists a one-to-one mapping α from $\mathscr{S}$ into $\mathscr{T}$ which satisfies the following properties:*

1) $\alpha(s_1 + s_2) = \alpha(s_1) + \alpha(s_2)$ for all $s_1, s_2 \in \mathscr{S}$.
2) $\alpha(s_1 \cdot s_2) = \alpha(s_1) \cdot \alpha(s_2)$ for all $s_1, s_2 \in \mathscr{S}$.
3) If $s_1 < s_2$ for $s_1, s_2 \in \mathscr{S}$, then $\alpha(s_1) < \alpha(s_2)$.

Let us examine these conditions. In (1) the addition on the left is the addition in $\mathscr{S}$ and the addition on the right is the addition in $\mathscr{T}$. Hence (1) says that we get the same result when we add s_1 and s_2 in $\mathscr{S}$ and then apply α or when we apply α to s_1 and to s_2 and add the images in $\mathscr{T}$. A mapping satisfying (1) is said to *preserve the additive structure* and is often called a *homomorphism* (meaning "same form") in a modern algebra course. Similar remarks can be made concerning (2) and (3).

When $(\mathscr{S}, +, \cdot, <)$ is embedded in $(\mathscr{T}, +, \cdot, <)$ by a mapping α, then α is a one-to-one mapping from $\mathscr{S}$ *onto* a subset of $\mathscr{T}$, namely, $\mathscr{U} = \{t \in \mathscr{T} \mid \alpha(s) = t \text{ for some } s \in \mathscr{S}\}$. Thus $\mathscr{U}$ and $\mathscr{S}$ are essentially the sets; they have the same cardinality, and conditions (1), (2), and (3) guarantee that the structures of $\mathscr{U}$ and $\mathscr{S}$ are the same. The difference between $\mathscr{U}$ and $\mathscr{S}$ is simply that the elements have "different names." Hence mathematicians often *identify* an element $s \in \mathscr{S}$ and the element $\alpha(s) \in \mathscr{T}$; i.e., rather than calling them different names, they relabel the elements of $\mathscr{S}$ and use the names of the elements of $\mathscr{T}$. This process will be clarified when we embed $(\mathscr{N}, +, \cdot, <)$ in $(\mathscr{I}, +, \cdot, <)$.

To carry out this embedding, we must first define a mapping α from $\mathscr{N}$ to $\mathscr{I}$. For a natural number $n \in \mathscr{N}$, define $\alpha(n) = [n^*, 1]$. We must now show that α is one-to-one and that (1), (2), and (3) hold.

Suppose $\alpha(n) = \alpha(m)$ for $n, m \in \mathscr{N}$; that is, $[n^*, 1] = [m^*, 1]$. Then $m^* + 1 = n^* + 1$. By the Cancellation Law of Addition for natural numbers, $m^* = n^*$. Since the mapping $*$ is one-to-one, $m = n$. Hence α is one-to-one.

We will verify that (2) and (3) hold and leave the verification of (1) as an exercise. Let $m, n \in \mathscr{N}$. Then

$$\begin{aligned}\alpha(m \cdot n) &= [(mn)^*, 1] \\ &= [mn + 1 + (m + n + 1), 1 + (m + n + 1)] \\ &= [(m + 1)(n + 1) + 1, (m + 1) + (n + 1)] \\ &= [m^*n^* + 1, m^* + n^*] \\ &= [m^*, 1] \cdot [n^*, 1] \\ &= \alpha(m) \cdot \alpha(n).\end{aligned}$$

In the second line of this argument we added $m + n + 1$ to both members of the equivalence class. To see that this is valid, simply note that the ordered pairs $(mn + 1, 1)$ and

$$\big(mn + 1 + (m + n + 1), 1 + (m + n + 1)\big)$$

are equivalent under the relation defined in 3.2.1.

Now suppose $m, n \in \mathscr{N}$ and $m < n$. We want to show that $\alpha(m) < \alpha(n)$; i.e., that $[m^*, 1] < [n^*, 1]$, which is equivalent to showing that $m^* + 1 < n^* + 1$ by Definition 3.2.5. Since $m < n$, there exists $p \in \mathscr{N}$ such that $p + m = n$. Thus $n^* + 1 = (p + m)^* + 1 = p + m^* + 1$ and hence $m^* + 1 < n^* + 1$, which is what we wanted to show.

Therefore α is an embedding of $(\mathscr{N}, +, \cdot, <)$ into $(\mathscr{I}, +, \cdot, <)$. Henceforth then we shall identify the natural number n with the integer $[n^*, 1]$ and by doing so we shall consider $\mathscr{N}$ to be a subset of $\mathscr{I}$. Also this proves that $+$, $\cdot$, and $<$ on $\mathscr{N}$ are exactly the same as $+$, $\cdot$, and $<$ on $\mathscr{I}$.

3.2.8 Exercises

1. Prove Theorem 3.2.2.
2. Prove that $\cdot$ as defined on $\mathscr{I}$ is well-defined.
3. a) Prove part (2) of Theorem 3.2.4.
   b) Prove that for $n, m \in \mathscr{N}$, $[n, n] = [m, m]$ and that this is the additive identity of $\mathscr{I}$.
   c) For each $a \in \mathscr{I}$, find $-a \in \mathscr{I}$ satisfying part (4) of Theorem 3.2.4.
   d) Prove part (5) of Theorem 3.2.4.
   e) For $n, m \in \mathscr{N}$, prove that $[n^*, n] = [m^*, m]$ and that this is the multiplicative identity of $\mathscr{I}$.
4. Prove that $3 < 5$ and $-2 < 3$ for *integers* -2, 3, and 5.
5. Prove parts (1), (3), and (4) of Theorem 3.2.6.
6. Use parts (1), (3), and (4) of Theorem 3.2.6 to prove the Cancellation Law of Multiplication for $\mathscr{I}$.
7. Complete the verification that $(\mathscr{N}, +, \cdot, <)$ is embedded in $(\mathscr{I}, +, \cdot, <)$.
8. Prove that the additive and multiplicative identities of $\mathscr{I}$ guaranteed by Theorem 3.2.4 are unique.
9. Do all elements of $\mathscr{I}$ have multiplicative inverses? If so, prove it. If not, give examples.

3.3 THE RATIONAL NUMBERS

So far we have constructed $(\mathscr{N}, +, \cdot, <)$ and $(\mathscr{I}, +, \cdot, <)$. The system of integers overcame one defect of $\mathscr{N}$, namely, the failure of all elements to have additive inverses. But one can easily see that $\mathscr{I}$ has at least one defect. There are elements of $\mathscr{I}$ which have no multiplicative inverse in $\mathscr{I}$. We cannot expect that zero will have a multiplicative inverse. But we do desire a system in which all nonzero elements have a multiplicative inverse.

This system is the system of rational numbers. We construct them now using only $\mathscr{I}$ and $\mathscr{N}$ and the properties we have proved about them.

From our past experience with rational numbers, we expect them to be "fractions" a/b, where $a, b \in \mathscr{I}$. Also we realize that b cannot be zero. Furthermore, we want a special kind of equality; for instance, we want $\frac{6}{8} = \frac{3}{4}$.

A first guess as to the actual construction of the set of rational numbers would be to take all ordered pairs (a, b), where $a, b \in \mathscr{I}$ and $b \neq 0$ as the fractions. But since $(6, 8) \neq (3, 4)$, we would not have the correct concept of equality. Hence we must employ an equivalence relation on the collection of ordered pairs which will put ordered pairs like $(6, 8)$ and $(3, 4)$ in the same equivalence class; i.e., we must have the ordered pairs (a, b) and (ax, bx) in the same equivalence class for all $a, b, x \in \mathscr{I}$, $b \neq 0$, $x \neq 0$.

3.3.1 Definition. *Let $\mathscr{P} = \{(a, b) \mid a, b \in \mathscr{I}, b \neq 0\}$. Define $\sim$ on $\mathscr{P}$ by $(a, b) \sim (c, d)$ if and only if $ad = bc$ for all $(a, b), (c, d) \in \mathscr{P}$.*

3.3.2 Theorem. *The relation $\sim$ on $\mathscr{P}$ is an equivalence relation.*

Proof. Let (a, b), (c, d), and $(f, g) \in \mathscr{P}$.
1) Since $ab = ab$, we have $(a, b) \sim (a, b)$. Hence $\sim$ is reflexive.
2) Suppose $(a, b) \sim (c, d)$. Then $ad = bc$. Thus since multiplication on $\mathscr{I}$ is commutative, $da = cb$. Hence $(c, d) \sim (a, b)$ and so $\sim$ is symmetric.
3) Suppose $(a, b) \sim (c, d)$ and $(c, d) \sim (f, g)$. Then $ad = bc$ and $cg = df$. Multiplying the first equation by g and the second by b, we get $adg = bcg$ and $cgb = dfb$. From these last two equations, we get $adg = dfb$, that is, $(ag)d = (bf)d$. Since $(c, d) \in \mathscr{P}$, $d \neq 0$. Since the Cancellation Law of Multiplication holds in $\mathscr{I}$ by 3.2.8, Exercise 6, d can be cancelled to give $ag = bg$. Hence $(a, b) \sim (f, g)$, proving that $\sim$ is transitive. Hence $\sim$ is an equivalence relation.

We denote the equivalence classes of $(a, b) \in \mathscr{P}$ by $[a, b]$ rather than by $[(a, b)]$. Then

$$\begin{aligned}[a, b] &= \{(c, d) \in \mathscr{P} \mid (c, d) \sim (a, b)\} \\ &= \{(c, d) \in \mathscr{P} \mid ad = bc\}.\end{aligned}$$

Therefore two equivalence classes $[a, b]$ and $[c, d]$ are equal if $(a, b) \sim (c, d)$, that is, if $ad = bc$. Hence $[a, b] = [ax, bx]$ for all nonzero $x \in \mathscr{I}$. Thus if one considers the equivalence classes $[a, b]$ as the "fractions" described above, then they have the property of equality which was insisted on in the heuristic description, namely, $[a, b] = [ax, bx]$ for all $(a, b) \in \mathscr{P}$ and all nonzero $x \in \mathscr{I}$.

3.3.3 Definition. *Let $\mathscr{Q}$ be the set of all equivalence classes of $\sim$ on $\mathscr{P}$. Define $+$ and $\cdot$ on $\mathscr{Q}$ by*

$$[a, b] + [c, d] = [ad + bc, bd]$$

and

$$[a, b]\cdot[c, d] = [ac, bd]$$

for all $[a, b], [c, d] \in \mathscr{Q}$. *The set is called the set of* ***rational numbers.***

As before when we have defined operations on equivalence classes, we must prove that these operations are well-defined.

3.3.4 Theorem. *The addition and multiplication defined above are well-defined.*

Proof. We consider the addition and leave the multiplication for an exercise. Suppose that $[a, b] = [a_1, b_1]$ and $[c, d] = [c_1, d_1]$. Then $ab_1 = ba_1$ and $cd_1 = dc_1$. Multiplying the first equation by dd_1 and the second by bb_1, we get

$$adb_1d_1 = a_1d_1bd \tag{1}$$

and

$$bcb_1d_1 = b_1c_1bd. \tag{2}$$

Adding (1) and (2) we get

$$(ad + bc)b_1d_1 = (a_1d_1 + b_1c_1)bd,$$

proving that

$$[ad + bc, bd] = [a_1d_1 + b_1c_1, b_1d_1].$$

Hence

$$[a, b] + [c, d] = [a_1, b_1] + [c_1, d_1].$$

3.3.5 Theorem. *The following properties hold for all* $r, s, t \in \mathscr{Q}$:

1) $r + (s + t) = (r + s) + t.$
2) $r + s = s + r.$
3) There exists $u \in \mathscr{Q}$ *such that* $r + u = r$ *for all* $r \in \mathscr{Q}$.
4) For each $r \in \mathscr{Q}$ *there exists* $v \in \mathscr{Q}$ *such that* $r + v = u$.
5) $r\cdot(s\cdot t) = (r\cdot s)\cdot t.$
6) $r\cdot s = s\cdot r.$
7) There exists $w \in \mathscr{Q}$ *such that* $r\cdot w = r$ *for all* $r \in \mathscr{Q}$.
8) For each $r \in \mathscr{Q}$, $r \neq u$, *there exists* $x \in \mathscr{Q}$ *such that* $r\cdot x = w$.
9) $r(s + t) = rs + rt.$

We leave the proofs of all but (4) as exercises.

Proof of (4). The additive identity $u \in \mathscr{Q}$ is given by $[0, a]$ for any nonzero $a = \mathscr{I}$. Note that $[0, a] = [0, b]$ for any nonzero $a, b \in \mathscr{I}$, so the additive identity does not depend on a or b. To see that $[0, a]$ is the additive identity of $\mathcal{Q}$, let $r = [b, c]$ for $b, c \in \mathcal{I}$, $c \neq 0$. Then

$$\begin{aligned}[b, c] + [0, a] &= [ba + 0, ca] \\ &= [ba, ca] \\ &= [b, c].\end{aligned}$$

Now let us simplify our notation. We have established the set $\mathcal{Q}$ of rational numbers as the set of all equivalence classes $[a, b]$, where $a, b \in \mathcal{I}$, $b \neq 0$. Let us use the symbol a/b to represent this equivalence class. Then our definition of addition and multiplication can be expressed by

$$\frac{a}{b} + \frac{c}{d} = \frac{ad + bc}{bd}$$

and

$$\frac{a}{b} \cdot \frac{c}{d} = \frac{ac}{bd}$$

for all a/b, $c/d \in \mathcal{Q}$. Moreover, our definition of equivalence relation gives us the fact that

$$\frac{a}{b} = \frac{ax}{bx}$$

for all $a/b \in \mathcal{Q}$ and all nonzero $x \in \mathcal{I}$.

Note that $(-a)/b = a/(-b)$ since $(-a)(-b) = ab$. Hence every rational number can be expressed as a/b, where $a, b \in \mathcal{I}$ and $b > 0$. Moreover, if a and b have a common factor c such that $a = ca_1$ and $b = cb_1$, then $a/b = ca_1/cb_1 = a_1/b_1$. Consequently, every rational number can be expressed as a/b, where $a, b \in \mathcal{I}$, $b > 0$, and a and b have no common factors other than 1. When rational numbers are expressed in this manner, one says that they are in *reduced form.*

3.3.6 Definition. *For a/b, $c/d \in \mathcal{Q}$, we say that $a/b < c/d$ if and only if $ad < bc$.*

The verification that this ordering is well-defined and the proof of its properties are left as an exercise. Two additional properties of the system $(\mathcal{Q}, +, \cdot, <)$ of rational numbers are made specific by the next two theorems.

3.3.7 Theorem (Density Property). *Between any two distinct rational numbers, there is another rational number. In particular, if u and v are rational numbers such that $u < v$, then*

$$u < \frac{u + v}{2} < v.$$

Proof. Since $u < v$, $u + u < u + v$ by 3.3.11, Exercise 4. Hence $2u < u + v$. By multiplying both sides by the positive rational number $1/2$, we get

$$u < \frac{u + v}{2}.$$

Similarly, one can show that

$$\frac{u + v}{2} < v,$$

which completes the proof of the theorem.

Can we be sure that $(u + v)/2$ in the above proof is a rational number? The theorem asserts that it is, so one really should check on it. We leave it as an easy exercise.

3.3.8 Theorem (Archimedean Property). *If u and v are any positive rational numbers, then there exists a positive integer n such that $nu > v$.*

Proof. Let $u = a/b$ and $v = c/d$ where a, b, c, and d are all positive integers. Let $n = 2bc$. We must show that $2bc(a/b) > c/d$. But this is true if $2ad > 1$; and $a > 0$ and $d > 0$ clearly imply that $2ad > 1$.

In conclusion we give an example of a "number" which is *not* a rational number. We usually use $\sqrt{2}$ to denote the positive "number" whose square is 2; i.e., if $x = \sqrt{2}$, then $x^2 = 2$. To prove that $\sqrt{2}$ is not a rational number is a simple matter and the proof depends only upon simple divisibility properties of the system of integers. The proof is given in Chapter 4, Theorem 4.3.7, and so is not given here. The proof in Chapter 4 is readable and requires little more than a rudimentary knowledge of divisibility properties. The reader is urged to read it now.

We usually consider $\mathscr{I}$ as a subset of $\mathscr{Q}$, and we now proceed to embed the number system $(\mathscr{I}, +, \cdot, <)$ in $(\mathscr{Q}, +, \cdot, <)$. Intuitively this is accomplished by identifying an integer n with the fraction $n/1$.

3.3.9 Definition. *Define a mapping α from $\mathscr{I}$ into $\mathscr{Q}$ by $\alpha(n) = n/1$ for all $n \in \mathscr{I}$.*

3.3.10 Theorem. *This mapping $\alpha: \mathscr{I} \to \mathscr{Q}$ is an embedding of the number system $(\mathscr{I}, +, \cdot, <)$ into the number system $(\mathscr{Q}, +, \cdot, <)$.*

We will verify that α satisfies condition (1) of Definition 3.2.7 and leave the rest of the proof of the theorem as an exercise. Do not forget that α must be one-to-one. Let $m, n \in \mathscr{I}$. Then

$$\begin{aligned}\alpha(m + n) &= (m + n)/1 \\ &= m/1 + n/1 \\ &= \alpha(m) + \alpha(n).\end{aligned}$$

3.3.11 Exercises

1. What would be wrong with defining an addition $\oplus$ on $\mathscr{Q}$ by

$$\frac{a}{b} \oplus \frac{c}{d} = \frac{a + c}{b + d}$$

   for all $a/b, c/d \in \mathscr{Q}$? Explain in detail.

2. a) Prove that the multiplication defined in 3.3.3 is well-defined.
   b) Prove that the ordering defined in 3.3.6 is well-defined.

3. Prove Theorem 3.3.5.

4. In the proof of part (4) of Theorem 3.3.5, it was established that the additive identity of $\mathscr{Q}$ is $[0, a]$ for any nonzero $a \in \mathscr{I}$. Does this contradict the uniqueness of the additive identity? Why or why not?

5. a) Prove that the sum and product of positive rational numbers are positive rational numbers. (Positive means greater than additive identity.)
   b) Prove the trichotomy law for $\mathscr{Q}$.
   c) Prove the following statements for $r, s, t \in \mathscr{Q}$.

      i) If $r > s$, then $r + t > s + t$.
      ii) If $r > s$ and $t > 0$, then $rt > st$.
      iii) If $r > s$ and $t < 0$, then $rt < st$.
      iv) If $r > s$ and $s > t$, then $r > t$.

6. If r and s are positive rational numbers such that $r < s$, prove that $1/s < 1/r$.

7. If r and s are rational numbers such that $r < s$ and if u and v are positive rational numbers, prove that

$$r < \frac{ur + vs}{u + v} < s.$$

8. If r and s are rational numbers such that $r < s$ and if n is any integer, show that there exist rational numbers $t_1, t_2, \ldots, t_n$ such that

$$r < t_1 < t_2 < \cdots < t_n < s.$$

9. Prove or disprove: The set $\mathscr{Q}_p$ of positive rationals with the usual ordering is well-ordered. That is, prove or disprove that every nonempty subset of $\mathscr{Q}_p$ has a least element.

10. Those readers who will teach mathematics someday will probably be asked questions like "Why can't we have 3/0 in the rationals?" or "Why is 0/0 not equal to 1?" What do you reply? Prove that $a/0$ and 0/0 cannot be defined and be consistent with the other properties of the rationals.

3.4 THE REAL NUMBERS

It is a common practice to envision the numbers which we use as being the points on a line. However, from Theorem 4.3.7 we know that there is no rational number b such that $b^2 = 2$. Hence the rational numbers do not "fill up" the number line. So there is a need for a larger, more comprehensive number system. This system is the system of real numbers.

We will construct the system $\mathcal{S}$ of positive real numbers; then the system of all real numbers is obtained from $\mathcal{S}$ exactly as the integers are obtained from the natural numbers, by considering equivalence classes which represent the differences of two positive real numbers. There are two common methods for constructing the system of positive real numbers from the system of rational numbers. One is the *Cauchy sequence* method, named for the French mathematician Augustin Cauchy (1789–1857). The other is the *Dedekind cut* method. Although other men made contributions to this theory, it is usually attributed to the German mathematicians George Cantor (1845–1918) and Richard Dedekind (1831–1916).

A Dedekind cut is a special kind of subset of the set $\mathcal{Q}_p$ of positive rational numbers. These Dedekind cuts will correspond to the positive real numbers.

3.4.1 Definition. *A subset D of $\mathcal{Q}_p$ is a* **Dedekind cut** *if the following hold:*

1) $D \neq \varnothing$ and $D \neq \mathcal{Q}_p$.
2) If $d \in D$ and a is an arbitrary element of $\mathcal{Q}_p$ such that $a < d$, then $a \in D$.
3) If $d \in D$, then there exists $c \in D$ such that $d < c$.

Condition (2) states that if a Dedekind cut contains a positive rational number r, then it contains all positive rationals less than r. Condition (3) states that a Dedekind cut has no largest element. We will often shorten "Dedekind cut" to "cut."

As an example of a Dedekind cut, let $a \in \mathcal{Q}_p$ and define

$$D_a = \{b \in \mathcal{Q}_p \mid b < a\}.$$

Then it is a routine matter to verify that the three conditions of Definition 3.4.1 hold. Hence D_a is a cut for all $a \in \mathcal{Q}_p$. This shows that to every positive rational number there corresponds a Dedekind cut.

Another example of a Dedekind cut is provided by setting $D = \{a \in \mathcal{Q}_p \mid a^2 < 2\}$. Let us verify that D is in fact a Dedekind cut. Clearly, $1 \in D$ and $2 \notin D$, so that (1) holds. Now let $a \in D$ and assume $b < a$ for $a, b \in \mathcal{Q}_p$. Then $b < a$ implies that $b^2 < ab < a^2$ so that $b^2 < 2$. Hence $b \in D$ and (2) holds. Since $2 - a^2 > 0$, by Theorem 3.3.8, there exists a positive integer n such that $2 - a^2 > (2a + 1)/n$. Then

$$\begin{aligned}\left(a + \frac{1}{n}\right)^2 &= a^2 + \frac{2a}{n} + \frac{1}{n^2} \\ &\le a^2 + \frac{2a + 1}{n} \\ &< 2.\end{aligned}$$

Hence $a + 1/n \in D$ and $a < a + 1/n$. Hence (3) holds. In the sequel we will want to associate this cut D with the "number" $\sqrt{2}$.

3.4.2 Definition. *Let $\mathcal{S}$ denote the set of all Dedekind cuts. For all $C, D \in \mathcal{S}$, define*

$$C + D = \{c + d \mid c \in C, d \in D\}$$

and

$$C \cdot D = \{c \cdot d \mid c \in C, d \in D\}.$$

If $D \in \mathcal{S}$ and $\tilde{D}$ denotes the complement of D in $\mathcal{Q}_p$, set

$$D^{-1} = \{b \in \mathcal{Q}_p \mid b < d^{-1} \text{ for some } d \in \tilde{D}\}.$$

Then D^{-1} is called the ***inverse*** *of D.*

It is not obvious that $C + D$, $C \cdot D$, and D^{-1} are cuts for all $C, D \in \mathcal{S}$. We will prove that $C + D$ is a cut and leave the proof that $C \cdot D$ and D^{-1} are Dedekind cuts as an exercise.

3.4.3 Theorem. *For all Dedekind cuts C and D, $C + D$, $C \cdot D$, and D^{-1} are also Dedekind cuts.*

Proof.

1) Clearly $C + D \neq \varnothing$. Since $C \neq \mathcal{Q}_p$ and $D \neq \mathcal{Q}_p$, there exist $c, d \in \mathcal{Q}_p$ such that $c \notin C$, $d \notin D$. Thus $c + d$ is greater than every element in $C + D$, so $c + d \notin C + D$. Thus $C + D \neq \mathcal{Q}_p$.

2) Let $c + d \in C + D$ with $c \in C$, $d \in D$ and let $a \in \mathcal{Q}_p$ be such that $a < c + d$. We must show that $a \in C + D$. Note that $a/(c + d) < 1$ so $ac/(c + d) < c$. Hence $ac/(c + d) \in C$. Similarly $ad/(c + d) \in D$. Thus $a = ac/(c + d) + ad/(c + d) \in C + D$.
3) Again, let $c + d \in C + D$ with $c \in C$, $d \in D$. Since D is a cut, there exists $d_1 \in D$ such that $d < d_1$. Then $c + d_1 \in C + D$ and $c + d < c + d_1$ (by 3.3.4).

3.4.4 Theorem. *The following hold for all B, C, and $D \in \mathcal{S}$:*

1) $(B + C) + D = B + (C + D)$.
2) $B + C = C + B$.
3) $(B \cdot C) \cdot D = B \cdot (C \cdot D)$.
4) $B \cdot C = C \cdot B$.
5) $D_1 \cdot D = D$ *for all* $D \in \mathcal{S}$.
6) For each $D \in \mathcal{S}$, $D \cdot D^{-1} = D_1$.

The proofs of (1) through (4) are straightforward and are left as exercises along with (5).

Proof of (6). We will show the two sets $D \cdot D^{-1}$ and D_1 are equal in the usual way.

Let $a \in D \cdot D^{-1}$; then $a = bc$ where $b \in D$, $c \in D^{-1}$. Since $c \in D^{-1}$, there exists $d \in \tilde{D}$ such that $c < d^{-1}$. Then $dc < 1$. Since $b \in D$ and $d \in \tilde{D}$, $b < d$. Hence $bc < dc < 1$; therefore $a < 1$ so that $a \in D_1$. This proves that $D \cdot D^{-1} \subseteq D_1$.

Now let $a \in D_1$; then $a < 1$ and $a^{-1} > 1$. We will show that for some $d \in D$, $d^{-1}a \in D^{-1}$ and so $a = d(d^{-1}a) \in D \cdot D^{-1}$, which will complete the proof. There exists a positive integer n such that $1/n \in D$. Let $b = (1 - a)/an$. If $b \in \tilde{D}$, then there exists $c \in D$ such that $b + c \in \tilde{D}$. If $b \in D$, then there exists a smallest positive integer k such that $kb \in \tilde{D}$. Then $k > 1$, $(k - 1)b \in D$, and $(k - 1)b + b \in \tilde{D}$. In this case, let $c = (k - 1)b$. Then in either case, $c \in D$ and $b + c \in \tilde{D}$. Moreover, since $1/n \in D$, we can pick c such that $c > 1/n$. Then

$$\begin{aligned} 0 &< (cn - 1)(1 - a)/an \\ &= ca^{-1} - b - c. \end{aligned}$$

Hence $ca^{-1} \in \tilde{D}$. We can pick an element $d \in D$ such that $d > c$. Then $d^{-1}a < c^{-1}a = (ca^{-1})^{-1}$. Hence since $ca^{-1} \in \tilde{D}$, $d^{-1}a \in D^{-1}$. Thus, as claimed,

$$a = d(d^{-1}a) \in D \cdot D^{-1}.$$

3.4.5 Definition. *Let $C, D \in \mathscr{S}$. Define $C < D$ if and only if $C \subseteq D$.*

The verification that the usual properties of orderings hold is left as an exercise.

We now have the system of positive real numbers. If we carry out a construction on the positive real numbers as we did to obtain the system of integers from the natural numbers, we obtain the system of all ***real numbers,*** which we denote by $(\mathscr{R}, +, \cdot, <)$. Note that when the negative real numbers are included, the set is $\mathscr{R}$.

Next we must embed $(\mathscr{Q}, +, \cdot, <)$ into $(\mathscr{R}, +, \cdot, <)$. The mapping $\alpha: \mathscr{Q} \to \mathscr{R}$ given by $\alpha(x) = D_x$ for all $x \in \mathscr{Q}$ is the required mapping. The substance of 3.4.10, Exercise 6, is to verify this fact. Hence, as usual, we identify the elements of $\mathscr{Q}$ with their images under α in $\mathscr{R}$. Consequently, $\mathscr{Q} \subseteq \mathscr{R}$ and when we speak of a rational number, we may as well assume that it is also a real number. Moreover, even though by our construction the system of real numbers consists of Dedekind cuts for which we have used capital letters as symbols, it is quite common to use lower-case letters to denote elements of $\mathscr{R}$. Hence if we say "$x \in \mathscr{R}$," x will be a Dedekind cut. Therefore x is a set of positive rational numbers, and as such has elements itself. Hence the term "rational number" is used in two ways simultaneously: (1) as a real number, i.e., as a Dedekind cut of the form D_x, and (2) as an element of a Dedekind cut.

We have seen that the set of rational numbers and the set of real numbers have many properties in common. However, there is one important property which the set of real numbers enjoys that the set of rational numbers does not have. This property is *completeness.* We will now prove that $\mathscr{R}$ is complete, but that $\mathscr{Q}$ is not. But first we need some definitions and a theorem.

3.4.6 Definition. *Let F be any nonempty subset of $\mathscr{R}$. If there exists an element x such that $x \geq f$ for all $f \in F$, then x is called an* ***upper bound*** *of F.*

Let us consider some examples. Let $F = \{r \in \mathscr{R} \mid r \leq 4\}$. Then 4 and any real number larger than 4 are upper bounds of F. Hence upper bounds are not unique. If $E = \{r \in \mathscr{R} \mid r < 4\}$, then 4 and any real number larger than 4 are also upper bounds of E. Hence upper bounds need not belong to the set. On the other hand, not every set has an upper bound; for instance, the set of integers has no upper bound.

3.4.7 Definition. *Let F be a nonempty subset of $\mathscr{R}$. If the set of upper bounds of F has a least element y in it, then y is said to be the* ***least upper bound*** *of F.*

In both the sets E and F discussed above, 4 is the least upper bound. Note that the least upper bound need not belong to the set. We have seen that a set may not have any upper bounds. However, if a set of *real numbers* has an upper bound, then it has a least upper bound. This property is called the *completeness property*.

Before proving that $\mathscr{R}$ is complete, we will state and prove an important theorem.

3.4.8 Theorem (Dedekind). *Let U be a set of real numbers satisfying the following conditions.*

1) U contains a positive rational number and $U \neq \mathscr{R}$.
2) If $x \in U$ and $y < x$, then $y \in U$.

Then there exists a unique real number z such that $u < z$ implies $u \in U$ and $z < v$ implies $v \notin U$. Conditions (1) and (2) are known as the ***Dedekind Conditions.***

Proof. First we will show the uniqueness of z. Suppose there are two elements z_1 and z_2 satisfying the conclusion of the theorem. Then without loss of generality, we may assume that $z_1 < z_2$. Hence by 3.4.10, Exercise 8, there exists a rational number a such that $z_1 < a < z_2$. From $z_1 < a$, we see that $a \notin U$. On the other hand, from $a < z_2$, $a \in U$. This contradiction establishes the uniqueness of z.

Now we prove the existence of z. Define

$$Z = \{b \in \mathscr{Q}_p \mid b \in x \text{ for some real number } x \in U\}.$$

First we will show that Z is a Dedekind cut.

1) Since U contains a positive rational number, Z is not empty. Let y be a real number not in U and let c be a positive rational number not in y. Then $c \notin Z$, for if $c \in Z$, then $c \in x$ for some real number $x \in U$. But $x < y$, a contradiction.

2) Let a be an arbitrary element of $\mathscr{Q}_p$ such that $a < b$ for some $b \in Z$. Since $b \in x$ for some $x \in U$, then $a < b$ implies that $a \in x$. Hence $a \in Z$.

3) Let $b \in Z$. Then there exists a real number $x \in U$ such that $b \in x$. Then there exists $c \in x$ such that $b < c$. Hence Z is a Dedekind cut. We will use the lower-case z in place of Z for the remainder of the proof.

Now we must show that z satisfies the conclusion of the theorem. Suppose $u < z$. Then there exists $a \in \mathscr{Q}_p$ such that $a \notin u$, $a \in z$. But since $a \in z$, $a \in x$ for some $x \in U$. The facts $a \notin u$ and $a \in x$ imply that $u < x$. Hence since $x \in U$, (2) implies that $u \in U$. On the other hand, suppose $z < v$. Then there exists $c \in \mathcal{Q}_p$ such that $c \notin z$, $c \in v$. If $v \in U$, then $c \in z$, contrary to our choice of c. Hence $v \notin U$. This proves the theorem.

Condition (1) may be relaxed to require only that U contain a real number and $U \neq \mathcal{R}$ by realizing that every real number contains a rational number and, in case U contains no positive rationals, by using the construction of $\mathcal{R}$ from $\mathcal{S}$ to obtain the result. We use Theorem 3.4.8 in this relaxed sense in considering the theorem for which we have been preparing.

3.4.9 Theorem (Completeness Property). *The system $(\mathcal{R}, +, \cdot, <)$ of real numbers is complete; i.e., every nonempty subset of $\mathcal{R}$ which has an upper bound has a least upper bound.*

Proof. Let U be a nonempty subset of $\mathcal{R}$ which has an upper bound. We will determine the least upper bound of U. Define a subset of $\mathcal{R}$ by

$$V = \{x \in \mathcal{R} \mid \text{there exists } u \in U \text{ such that } u > x\}.$$

Next we prove that V satisfies the Dedekind Conditions.

1) Since U is not empty, there exists a real number $u \in U$. Then every real number less than u belongs to V, so V is not empty. Also since U has an upper bound, say y, there exist real numbers larger than y which are not elements of V. Thus $V \neq \mathcal{R}$.

2) Let $x \in V$ and suppose $y < x$. Since $x \in V$, there exists $u \in U$ such that $x < u$. Then $y < u$, so that $y \in V$.

Hence by Theorem 3.4.8, there exists a unique real number z such that $u < z$ implies $u \in V$ and $z < v$ implies that $v \notin V$. We will prove that z is the least upper bound of U. First we prove that z is an upper bound of U, and then that z is the *least* upper bound.

Suppose z is not an upper bound of U. Then there exists $r \in U$ such that $z < r$. Let s be a real number such that $z < s < r$. Since s is less than r, $s \in V$ by definition of V; but $z < s$ implies that $s \notin V$. This contradiction implies that z is an upper bound of U. Furthermore, for every positive real number t, $z - t < z$. Hence for every positive real number t, $z - t \in V$. Thus $z - t < u$ for some $u \in U$. Thus no real numbers less than z are upper bounds of U. Hence z is the least upper bound of U.

To show that the system $(\mathcal{Q}, +, \cdot, <)$ of rational numbers is not complete, we will show that the nonempty subset

$$A = \{x \in \mathcal{Q}_p \mid x^2 < 2\}$$

which has an upper bound does not have a least upper bound. Suppose A does have a least upper bound in $\mathcal{Q}$; call it z. Then $z^2 \in \mathcal{Q}$ and by the trichotomy law, exactly one of the following is true: $z^2 = 2$, $z^2 < 2$, or $z^2 > 2$. But there is no rational number such that $z^2 = 2$, so $z^2 < 2$ or $z^2 > 2$. We will show that these two possibilities lead to contradictions, and therefore will prove that A has no least upper bound in $\mathcal{Q}$.

If $z^2 < 2$, then one can show that if $x = 4z/(z^2 + 2)$, then $z < x$ and $x^2 < 2$, contradicting the fact that z is the least upper bound of A. On the other hand, if $z^2 > 2$, let $y = (z^2 + 2)/2z$; then $z > y$ and $y^2 > 2$. Hence for any $u \in A$, $u \leq y$, so y is an upper bound of A. Therefore $y < z$ contradicts our choice of z.

3.4.10 Exercises

1. Prove that for all $a \in \mathcal{Q}_p$, D_a is a Dedekind cut.
2. If C and D are Dedekind cuts, prove that $C \cdot D$ and D^{-1} are Dedekind cuts.
3. Prove parts (1) through (5) of Theorem 3.4.4.
4. Prove the trichotomy law for $<$ on $\mathcal{Q}$. (See Definition 3.4.5.)
5. Prove that for B, C, and D in $\mathcal{S}$, $B < C$ if and only if $B + D < C + D$.
6. Let $x, y \in \mathcal{Q}_p$. Prove the following:
   a) $D_x + D_y = D_{x+y}$.
   b) $D_x \cdot D_y = D_{xy}$.
   c) $D_x < D_y$ if and only if $x < y$.
7. Carry out the details of constructing $\mathcal{R}$ from $\mathcal{S}$.
8. Let C and D be arbitrary real numbers such that $C < D$. Prove that there exists a rational number a such that $C < D_a < D$.
9. Fill in all details of the discussion which proves that $\mathcal{Q}$ is not complete.
10. Write the number 2 as a natural number, as an integer, as a rational number, and as a real number. Use only natural numbers to denote these various representations of 2.

3.5 THE COMPLEX NUMBERS

The construction of the system of complex numbers from the system of real numbers is relatively simple. In this section we sketch the construction and leave the proofs as exercises. Throughout the section we will use lower-case letters to represent elements of $\mathcal{R}$.

From our experience in college algebra, we know that complex numbers are "numbers" of the form $a + bi$, where $i^2 = -1$. However, we cannot assume that such a number i exists. We avoid this by considering ordered pairs (a, b) of real numbers. However, one may intuitively think of (a, b) as $a + bi$; this will motivate the definitions we will give.

3.5.1 Definition. *Let $\mathscr{C}$ be the set of all ordered pairs (a, b) of real numbers. Define $+$ and $\cdot$ on $\mathscr{C}$ by*

$$(a, b) + (c, d) = (a + c, b + d)$$

and

$$(a, b)\cdot(c, d) = (ac - bd, ad + bc)$$

for all $(a, b), (c, d) \in \mathscr{C}$. Call $\mathscr{C}$ the set of ***complex numbers.***

3.5.2 Theorem. *The operations $+$ and $\cdot$ on $\mathscr{C}$ are commutative, associative binary operations. Furthermore, $\mathscr{C}$ has both an additive and a multiplicative identity, each element of $\mathscr{C}$ has an additive inverse, and each nonzero element of $\mathscr{C}$ has a multiplicative inverse.*

The proof is easy and is left as an exercise. Note that since we have not used equivalence classes, there is no question of well-definedness.

It is not possible to define an ordering on $\mathscr{C}$ which has the trichotomy property. Furthermore, we shall see below that there is a nonzero element of $\mathscr{C}$ whose square is negative. This also violates our ordering properties. Hence $\mathscr{C}$ fails to qualify as a number system in the sense of $\mathscr{N}$, $\mathscr{I}$, $\mathscr{Q}$, and $\mathscr{R}$.

To embed the real numbers in the complex numbers we define a mapping $\alpha: \mathscr{R} \to \mathscr{C}$ by $\alpha(a) = (a, 0)$. Here we verify that α is one-to-one and that conditions (1) and (2) of Definition 3.2.7 hold, since condition (3) does not make sense in $\mathscr{C}$. This proof is also easy. We use this mapping to identify the real number a and the complex number $(a, 0)$.

Note that any complex number (a, b) can be written in the form

$$(a, b) = (a, 0) + (b, 0)(0, 1).$$

Under the embedding $(a, 0)$ corresponds to a and $(b, 0)$ to b. If we use i for the complex number $(0, 1)$, then $(a, b) = a + bi$ and $i^2 = -1$. Hence the complex numbers have their usual form.

In this chapter we have progressed through a series of constructions which have netted us the usual number systems from a concrete basis. Furthermore, by identifying elements under the appropriate embeddings, we have established the following relationships among the sets involved:

$$\mathscr{N} \subseteq \mathscr{I} \subseteq \mathscr{Q} \subseteq \mathscr{R} \subseteq \mathscr{C}.$$

Note that at first as these sets increase, the corresponding systems gain important properties. For instance, $\mathcal{N}$ had no additive inverses, whereas the elements of $\mathcal{I}$ and higher systems do have. Elements in $\mathcal{I}$ had no multiplicative inverses, but the nonzero elements of $\mathcal{Q}$, $\mathcal{R}$, and $\mathcal{C}$ do have multiplicative inverses. Furthermore, $\mathcal{Q}$ is not complete, but $\mathcal{R}$ and $\mathcal{C}$ are. Hence each construction yielded a number system possessing a property that its predecessor did not have. However, in constructing $\mathcal{Q}$ from $\mathcal{I}$, the well-ordering property of the positive elements was lost. In fact $\mathcal{N}$ is well-ordered, but $\mathcal{I}$ itself is not. Also in passing from $\mathcal{R}$ to $\mathcal{C}$, all ordering properties were lost. Heuristically these facts may be explained this way: By increasing the algebraic properties of the systems, we have decreased the ordering properties.

3.5.3 Exercises

1. Prove Theorem 3.5.2.
2. Verify that $\alpha: \mathcal{R} \to \mathcal{C}$ satisfies conditions (1) and (2) of Definition 3.2.7.
3. The multiplication defined in 3.5.1 is somewhat cumbersome and, for those who are not familiar with the complex numbers, it is unnatural. What goes wrong if we take $\mathcal{C}$ to be the same set, define + as in 3.5.1, and take
$$(a, b)\cdot(c, d) = (ac, bd)$$
as our definition of multiplication?

ADDITIONAL REFERENCES FOR CHAPTER 3

Bartle, Robert G., *The Elements of Real Analysis*, New York: Wiley (1964).

Crouch, Ralph, and Elbert Walker, *Introduction to Modern Algebra and Analysis*, New York: Holt, Rinehart, and Winston (1962).

McCoy, Neal H., *Introduction to Modern Algebra*, Boston: Allyn and Bacon (1962).

Ribenboim, Paulo, *Functions, Limits, and Continuity*, New York: Wiley (1964).

Rudin, Walter, *Principles of Mathematical Analysis*, 2nd ed., New York: McGraw-Hill (1964).

Thurston, H. A., *The Number-System*, New York: Interscience (1956).

Youse, Bevan K., *The Number System*, Belmont, Calif.: Dickenson Publishing Company (1965).

4 properties of the integers

Throughout the following discussion it will be assumed that the reader has a general familiarity with the system $(\mathscr{I}, +, \cdot, <)$ of integers and certain properties which it possesses. This chapter presents a close look at the integers and establishes some rudimentary properties concerning divisibility, primes, the division algorithm, the greatest common divisor and the Euclidean algorithm, and the Fundamental Theorem of Arithmetic. Also the important property of the system of integers—mathematical induction—is discussed.

Chapter 3 contains a rigorous development of the system of integers and outlines proofs of the important properties of the integers which will be assumed to be "well known" for our purposes here. An effort has been made to keep this chapter independent of Chapter 3. The one major fact about the system of integers of which we will make frequent use is that the set of positive (and nonnegative) integers is *well-ordered*; i.e., every nonempty subset of the positive (nonnegative) integers has a least element. This fact is proved by Theorem 3.1.8 and the embedding of $\mathscr{N}$ into $\mathscr{I}$.

4.1 MATHEMATICAL INDUCTION

In this section we establish the Principle of Mathematical Induction, a principle which is the basis of proofs throughout mathematics. First, the following is essential.

4.1.1 Lemma. *Let Q be a set of positive integers with the following properties:*

1) $1 \in Q$.
2) If k is a positive integer such that $k \in Q$, then $k + 1 \in Q$.

Then Q is the set of all positive integers.

Proof. Assume that Q is not the set of all positive integers and let R be the set of all positive integers not in Q. Then since the positive integers are well-ordered, R has a smallest element; call it r. Then by (1), $r \neq 1$; hence $r - 1$ is a positive integer and $r - 1 \notin R$. Thus $r - 1 \in Q$. Applying (2) when $k = r - 1$ implies that $k + 1 = r \in Q$. But r was chosen as an element of R so that $r \notin Q$. This contradiction establishes the lemma.

4.1.2 Principle of Mathematical Induction. *Suppose that with each positive integer n there is associated a statement S_n. If*

1) S_1 is true, and

2) if k is a positive integer such that S_k is true, then S_{k+1} is true,

then S_n is true for all positive integers n.

Proof. Let Q be the set of positive integers n such that S_n is true. Then Q satisfies the hypotheses of Lemma 4.1.1. Therefore Q is the set of all positive integers; i.e., S_n is true for all positive integers n.

As an illustration of the use of the Principle of Mathematical Induction, let us consider a simple example from college algebra:

$$1 + 2 + \cdots + n = n(n + 1)/2$$

for all positive integers n. To prove this, let S_n be the statement "$1 + 2 + \cdots + n = n(n + 1)/2$." Then S_n says that the sum of the first n positive integers is $n(n + 1)/2$. S_1 simply states that $1 = (1 \cdot 2)/2$, so S_1 is true. Suppose that k is any positive integer such that S_k is true; i.e., suppose that $1 + 2 + \cdots + k = k(k + 1)/2$. By adding $k + 1$ to both sides, we get

$$\begin{aligned} 1 + 2 + \cdots + k + k + 1 &= k(k + 1)/2 + k + 1 \\ &= (k + 1)(k + 2)/2. \end{aligned}$$

Hence S_{k+1} is true. Thus by the Principle of Mathematical Induction, S_n is true for all positive integers n; hence the result holds.

Another application of the Principle of Mathematical Induction is in making *recursive* (or *inductive*) *definitions*. When one gives a definition which depends upon a positive integer n, the definition is often done recursively. For example, let a be any integer. We wish to define formally what is meant by a^n, where n can be any positive integer. First, define $a^1 = a$. Then if k is any positive integer such that a^k is defined, then define $a^{k+1} = a^k \cdot a$. If S_n is the statement "a^n is defined," then the Principle of Mathematical Induction shows that a^n is defined for all positive integers n.

As another illustration of the Principle of Mathematical Induction, we will prove that the "usual" rules of exponents are true under the definition given above. For instance, if a is an integer,

$$a^m \cdot a^n = a^{m+n}$$

for all positive integers m and n. Here we must prove a statement which is true for all values of the positive integers m and n. Hence we let S_n be the statement "$a^m a^n = a^{m+n}$ for all positive integers m." Thus S_n is a statement which involves all values of m, but only one value of n. S_1 is simply "$a^m a^1 = a^{m+1}$ for all positive integers m." This is true by the definition of a^{m+1}. Assume that S_k is true; i.e., assume that $a^m a^k = a^{m+k}$ for all positive integers m. Then for all positive integers m,

$$\begin{aligned} a^m a^{k+1} &= a^m(a^k a^1) && \text{(definition of } a^{k+1}\text{)},\\ &= (a^m a^k)a^1 && \text{(associative law)},\\ &= a^{(m+k)}a^1 && \text{(by } S_k\text{)},\\ &= a^{m+k+1} && \text{(definition of } a^{m+k+1}\text{)}. \end{aligned}$$

Hence S_{k+1} is true. Thus by the Principle of Mathematical Induction, S_n is true for all positive integers n. This proves the exponent law as asserted.

4.1.3 Exercises

1. Prove the following formulas from college algebra.

   a) $1 + 2^2 + 3^2 + \cdots + n^2 = n(n + 1)(n + 2)/6$
   b) $2 + 4 + 6 + \cdots + 2n = n(n + 1)$
   c) $3 + 3^2 + 3^3 + \cdots + 3^n = 3(3^n - 1)/2$
   d) $a + (a + d) + (a + 2d) + \cdots + (a + nd) = (n + 1)(2a + nd)/2$
   e) $a + ar + ar^2 + \cdots + ar^n = a(1 - r^{n+1})/(1 - r)$

2. Let x and y be integers and let m and n be positive integers. Prove the following rules for exponents. Use the definition given above.

   a) $(xy)^n = x^n y^n$ b) $(x^n)^m = x^{nm}$

3. a) Use the fact that the set of positive integers is well-ordered (every nonempty set of positive integers has a smallest element) to prove the following:

   Strong Principle of Mathematical Induction. Suppose that with each positive integer n there is associated a statement S_n. Suppose that

   1) S_1 is true, and
   2) if k is a positive integer such that S_i is true for all $i < k$, then S_k is true.

   Then S_n is true for all positive integers n.

b) Note that this principle of mathematical induction is actually equivalent to 4.1.2. Thus the "strong" in the name is misleading. Prove the equivalence of the two forms of the Principle of Mathematical Induction.

4. We know that if a_1 and a_2 are integers, then $(-a_1)(-a_2) = a_1a_2$ and $(a_1)(-a_2) = -(a_1a_2)$. Prove that if n is any positive integer and $a_1, a_2, \ldots, a_n$ are integers, then

$$(-a_1)(-a_2)\cdots(-a_n) = \begin{cases} a_1a_2\cdots a_n & \text{if } n \text{ is even,} \\ -(a_1a_2\cdots a_n) & \text{if } n \text{ is odd.} \end{cases}$$

4.2 DIVISIBILITY AND PRIMES

4.2.1 Definition. *Let $a, b \in \mathscr{I}$. Then we say a* ***divides*** *b if there exists $c \in \mathscr{I}$ such that $b = ac$. In this case we also say that a is a* ***divisor*** *of b, a is a* ***factor*** *of b, and b is a* ***multiple*** *of a. If a divides b, we denote this fact by $a \mid b$.*

The symbol $\nmid$ is used to denote "does not divide"; for example, $3 \nmid 4$. Note that for any integer b, ± 1 and $\pm b$ divide b. Moreover b divides 0 since $0 = b \cdot 0$. Also if a and b are positive integers and $a \mid b$, then $a \leq b$. To see this, observe that since $a \mid b$, there exists $c \in \mathscr{I}$ such that $b = ac$. Since a and b are positive, c must be positive. Hence $c \geq 1$, and if we multiply by a we get $b = ac \geq a$.

The symbol $\mid$ defines a relation on $\mathscr{I}$. We have seen above that $\mid$ is reflexive. But $\mid$ is not symmetric since $2 \mid 4$, but $4 \nmid 2$. Thus $\mid$ is not an equivalence relation on $\mathscr{I}$. However, $\mid$ is transitive; for, suppose $a \mid b$ and $b \mid c$. Then there exist $d, f \in \mathscr{I}$ such that $b = da$ and $c = fb$. Then $c = f(da) = (fd)a$ so that $a \mid c$.

Certain integers stand out because of their divisibility properties. These integers are called *primes*.

4.2.2. Definition. *A positive integer p is a* ***prime*** *if $p \neq 1$ and if the only divisors of p are ± 1 and $\pm p$.*

Thus by the definition, if p is a prime then $p \geq 2$. Some authors do not require that p be positive in order to be a prime. However, in order to obtain "uniqueness" results in the discussion of factorization in Section 4.3, this requirement is necessary. Requiring primes to be positive is not a serious restriction since "negative primes" can be expressed as -1 times a prime. It is worth while to consider the negation of Definition 4.2.2. Suppose $n \in I$ and n is not a prime. Then n is negative, or $n = 0$, or $n = 1$, or if none of these is true, then there exist integers n_1 and n_2 such that

$$n = n_1 n_2, \qquad 1 < n_1 < n, \qquad \text{and} \qquad 1 < n_2 < n.$$

Primes play an important role in the theory of numbers. In fact, they are the building blocks from which the integers arise. We will prove in Section 4.3 that every nonzero integer can be expressed uniquely as ± 1 times a product of a finite number of primes. Much of the elementary theory of numbers is devoted to a study of primes—their properties and existence. There are many unsolved problems in number theory dealing with primes. For more information on primes, the reader is referred to contemporary texts in elementary number theory. (See the list of references at the end of the chapter.)

Now let c and b be positive integers. We want to develop a process by which we can "divide" c by b. We must be careful here since there is no binary operation "division" defined on $\mathscr{I}$. Intuitively if we consider the number line and mark off multiples of b, then c must either coincide with one of the multiples of b or fall immediately to the right of one of them:

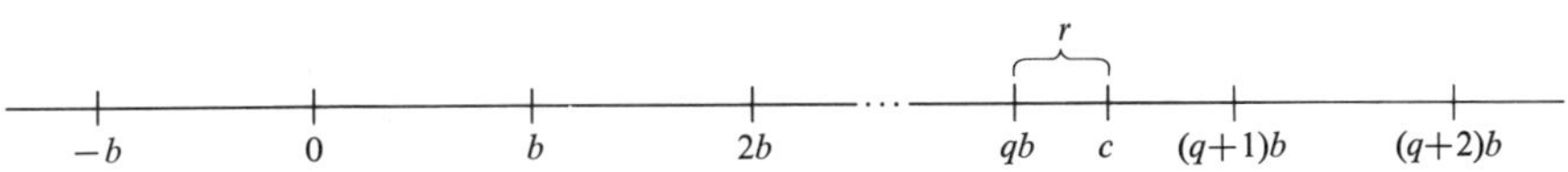

Suppose c coincides with or falls immediately to the right of qb. If we denote the length of the line segment from qb to c by r, then we can write $c = qb + r$. Moreover, if c is to the left of $(q + 1)b$, then $0 \leq r < b$. Hence we have "divided" b into c with "quotient" q and "remainder" r. For example, if we divide 67 by 14 we get $67 = 4(14) + 11$.

Now let us make this concept precise. The process we have described is called the *Division Algorithm*. The proof uses the fact that the set of nonnegative integers is well-ordered (see Section 2.6); that is, given any nonempty subset S of nonnegative integers, S must contain a smallest element.

4.2.3 Theorem (Division Algorithm). *Given any two positive integers c and b, there exist unique integers q and r such that*

(1) $$c = bq + r \qquad \textit{and} \qquad 0 \leq r < b.$$

Proof. First we must prove the existence of integers q and r satisfying (1) and then prove that they are unique. Let

$$S = \{a - bx \mid x \in \mathscr{I}, c - bx \geq 0\}.$$

Recall that c and b are fixed; so S consists of all nonnegative expressions of the form $c - bx$ as x varies in $\mathscr{I}$. For $x = 0$, $c - bx = c > 0$. Hence $c \in S$ so S is nonempty. Thus by the well-ordering property of the nonnegative integers, S contains a smallest element, say r. Then $r \in S$, so that there exists $q \in \mathscr{I}$ such that $c - bq = r$. Thus $c = bq + r$ and $0 \leq r$. To verify that q and r satisfy (1), we must now show that $r < b$. Suppose not; i.e., suppose $r \geq b$. Then $r - b \geq 0$ and $r - b = (c - bq) - b = c - (b + 1)q$. This proves that $r - b \in S$. But since b is positive, $r > r - b$. This contradicts the fact that r is the smallest element of S. Hence we must have $r < b$. This completes the proof of the existence of q and r.

Now suppose there exist integers q_1 and r_1 such that $c = bq_1 + r_1$ and $0 \leq r_1 < b$. Then $bq_1 + r_1 = bq + r$, and hence $b(q_1 - q) = r - r_1$. Without loss of generality we may assume that $q_1 \geq q$. (For if $q_1 < q$, consider $b(q - q_1) = r_1 - r$ and use the argument given here.) Then $q_1 - q \geq 0$. If $q_1 - q \neq 0$, then $q_1 - q \geq 1$, so that $r_1 - r = b(q - q_1) \geq b$, contradicting the fact that $b > r_1$ and $b > r$. Hence $q_1 - q = 0$ and consequently $r - r_1 = 0$. Therefore $q_1 = q$ and $r_1 = r$, establishing the uniqueness.

4.2.4 Remark

The hypothesis that b is a positive number can be relaxed to require that b be nonzero by noting that if b is negative, then $-b$ is positive. Hence $c = (-b)(-q) + r$. Furthermore, the restriction that c is positive can also be removed.

Given c and b, if q and r are the unique integers satisfying (1) in Theorem 4.2.3, we call r the *remainder* of dividing c by b and call q the *quotient*.

4.2.5 Exercises

Let a, b, c, d, p, and q be arbitrary integers.

1. For each pair of numbers given below, find the q and r which satisfy (1) of Theorem 4.2.3.

   a) $c = 162, b = 49$ b) $c = 213, b = 114$
   c) $c = 1032, b = -25$

2. If $a \mid b$ and $a \mid c$, prove that $a \mid (b \pm c)$.
3. If $a \mid b$, prove that $a \mid bc$ for all $c \in \mathscr{I}$.

4. Prove or disprove: If $a \mid bc$, then $a \mid b$ or $a \mid c$.
5. If $a \mid b$ and $b \mid a$, show that $a = \pm b$. In this case, a and b are said to be *associates*.
6. Suppose $a \mid b$. Prove or disprove the following statements.

   a) $ac \mid bc$ for all $c \in \mathscr{I}$.
   b) $a + c \mid b + c$ for all $c \in \mathscr{I}$.
   c) If in addition $d \mid b$, then $a + d \mid b$.

7. Let p and q be primes. If $p \mid q$, the $p = q$.
8. If p and q are primes, show that $p \nmid (pq + 1)$ and $q \nmid (pq + 1)$.
9. If n is a positive integer and $p_1, p_2, \ldots, p_n$ are distinct primes, show that $p_1 p_2 \cdots p_n + 1$ is not divisible by any of the primes $p_1, p_2, \ldots, p_n$.
10. Use Exercise 9 to prove that there are infinitely many primes.
11. If $a = b + c$ and if d divides any two of the three integers a, b, and c, prove that d must also divide the third.

4.3 GREATEST COMMON DIVISOR

4.3.1 Definition. *Let a and b be given nonzero integers. The positive integer d is called the* **greatest common divisor of a and b** *if the following conditions hold:*

1) $d \mid a$ and $d \mid b$.
2) If $e \mid a$ and $e \mid b$, then $e \mid d$.

Condition (1) states that d must be a common divisor of a and b, and condition (2) implies that d is the "greatest" common divisor. We require the greatest common divisor of two nonzero integers to be positive to obtain uniqueness. Note that if d is the greatest common divisor of a and b, then $-d$ satisfies conditions (1) and (2) also. So far we have no evidence that given two nonzero integers it is possible to find their greatest common divisor. We will prove that the greatest common divisor of any two nonzero integers exists; and, in fact, we will show that the greatest common divisor can be expressed in a special form. Then we will develop an algorithm by which we can determine the greatest common divisor of any two nonzero integers.

4.3.2 Theorem. *Let a and b be given nonzero integers. Then the greatest common divisor d of a and b exists; and, in fact, there exist integers m and n such that $d = am + bn$.*

Proof. Let $S = \{ax + by \mid x, y \in \mathscr{I}, ax + by > 0\}$. Then letting $x = a$ and $y = b$, we obtain $a^2 + b^2 \in S$, since $a^2 + b^2 > 0$. Hence S is not empty. Thus by the well-ordering of the positive integers, S has a least element. Let d be this least element; then $d = am + bn$ for some $m, n \in \mathscr{I}$. We will prove that d is the greatest common divisor of a and b.

First we prove that $d \mid a$. That $d \mid b$ is proved similarly. By the Division Algorithm, there exist integers q and r such that

$$a = qd + r \qquad \text{and} \qquad 0 \leq r < d.$$

Then $a = q(am + bn) + r$, so that $r = a(1 - qm) + b(qn)$. Thus if $r \neq 0$, then $r \in S$. But since $r < d$, this contradicts our choice of d. Hence $r = 0$. Thus $a = qd$, proving that $d \mid a$.

Now suppose $e \mid a$ and $e \mid b$ for some $e \in \mathscr{I}$. Then there exist $u, v \in \mathscr{I}$ such that $a = eu$ and $b = ev$. Then $d = (eu)m + (ev)n = e(um + vn)$. Therefore $e \mid d$. This proves that d is the greatest common divisor of a and b.

4.3.3 Definition. *Let a and b be given integers. Then an expression of the form $ax + by$ is called a* ***linear combination*** *of a and b.*

This definition allows us to restate Theorem 4.3.2 as the following corollary.

4.3.4 Corollary. *Let a and b be given nonzero integers. Then the greatest common divisor of a and b is the least positive linear combination of a and b.*

4.3.5 Definition. *Two nonzero integers are said to be* ***relatively prime*** *if their greatest common divisor is 1.*

Thus two nonzero integers are relatively prime if they have no common factors other than one. If p is a prime and a is a nonzero integer, then a and p are relatively prime unless a is a multiple of p. Thus primes are either relatively prime or equal. On the other hand, two nonzero integers can be relatively prime and neither be prime; for instance, 8 and 9 are relatively prime, but neither is prime.

Do not misinterpret Corollary 4.3.4! If a and b are nonzero integers such that $d = ax + by$ for some $d, x, y \in \mathscr{I}$, then Corollary 4.3.4 does not imply that d is the greatest common divisor of a and b. Indeed, d is the greatest common divisor of a and b only if d is the *least positive* linear combination of a and b. On the other hand, if a and b are nonzero integers such that $1 = ax + by$ for some $x, y \in \mathscr{I}$, then a and b are relatively prime. The greatest common divisor of a and b is

1, since 1 is clearly the least positive linear combination of a and b. Note that $1 = ax + by$ means that x and y, as well as a and y, are also relatively prime.

For convenience, the symbol (a, b) is usually used to denote the greatest common divisor of a and b. This symbol is not to be confused with the ordered pair (a, b). Thus a and b are relatively prime if $(a, b) = 1$. When confusion may arise from the use of this symbol, "greatest common divisor" is often shortened to g.c.d.

Theorem 4.3.2 established the existence of the g.c.d. of any two nonzero integers, but did not provide a method for finding the g.c.d. of two given nonzero integers. We will now discuss a procedure by which the g.c.d. of any two specific nonzero integers can be computed. This method is known as the *Euclidean Algorithm.*

Let a and b be given nonzero integers. Since $(a, b) = (-a, b)$ we may assume that a is positive without loss of generality. By the Division Algorithm (and 4.2.4) there exist integers q_1 and r_1 such that

$$a = bq_1 + r_1 \qquad \text{and} \qquad 0 \leq r_1 < b.$$

If $r_1 \neq 0$, divide b by r_1. Then there exist integers q_2 and r_2 such that

$$b = r_1 q_2 + r_2 \qquad \text{and} \qquad 0 \leq r_2 < r_1.$$

Continue this process as long as the remainder is nonzero. At each step, the divisor of the previous step is divided by the remainder of the same step, thus producing a new remainder and divisor. After a finite number of steps we will have the following:

$$\begin{array}{lrcll}
(1) & a &=& bq_1 + r_1 & \text{and} \quad 0 < r_1 < b, \\
(2) & b &=& r_1 q_2 + r_2 & \text{and} \quad 0 < r_2 < r_1, \\
(3) & r_1 &=& r_2 q_3 + r_3 & \text{and} \quad 0 < r_3 < r_2, \\
 & \vdots & & & \qquad\quad \vdots \\
(\mathrm{k}-1) & r_{k-3} &=& r_{k-2} q_{k-1} + r_{k-1} & \text{and} \quad 0 < r_{k-1} < r_{k-2}, \\
(\mathrm{k}) & r_{k-2} &=& r_{k-1} q_k + r_k & \text{and} \quad 0 < r_k < r_{k-1}, \\
(\mathrm{k}+1) & r_{k-1} &=& r_k q_{k+1} + 0. &
\end{array}$$

We must eventually obtain a remainder of 0, since $r_1, r_2, \ldots, r_k$ are positive integers such that

$$r_1 > r_2 > r_3 > \cdots > r_{k-1} > r_k.$$

4.3.6 Theorem (Euclidean Algorithm). *The last nonzero remainder in the process above is the g.c.d. of a and b.*

Proof. We must show that $r_k = (a, b)$. First we show that r_k divides a and b. From (k + 1) it is clear that $r_k \mid r_{k-1}$. Thus from (k) it follows that $r_k \mid r_{k-2}$ (since $r_k \mid r_{k-1}q_k$ and r_k divides itself) by 4.2.5, Exercise 2. Then from (k − 1), a similar argument shows that $r_k \mid r_{k-3}$. Continue! Finally we see that $r_k \mid r_1$ and $r_k \mid r_2$. Thus $r_k \mid b$ by (2). Then from (1) we get that $r_k \mid a$.

Now suppose $e \mid a$ and $e \mid b$. Then from (1), using 4.2.5, Exercise 2, we see that $e \mid r_1$. Similarly, from (2), $e \mid r_2$, and so on. Finally, by continuing through each step, we arrive at the conclusion $e \mid r_k$. This proves that r_k is the g.c.d. of a and b.

Let us illustrate by considering a numerical example. Find (422, 68).

$$\begin{aligned} 422 &= 68\cdot 6 + 14, \\ 68 &= 14\cdot 4 + 12, \\ 14 &= 12\cdot 1 + \ \,2, \\ 12 &= \ \,2\cdot 6 + \ \,0. \end{aligned}$$

Thus (422, 68) = 2.

Theorem 4.3.2 guaranteed that the g.c.d. of two nonzero integers a and b is expressible as a linear combination of a and b. The computation of the g.c.d. via the Euclidean Algorithm provides the equations necessary to find the correct linear combination. From (1) we see that

$$r_1 = a - bq_1.$$

Then combining this with (2), we get

$$\begin{aligned} r_2 &= b - r_1q_2 \\ &= b - (a - bq_1)q_2 \\ &= -q_2a + (1 + q_1q_2)b. \end{aligned}$$

Next,

$$\begin{aligned} r_3 &= r_1 - r_2q_3 \\ &= (a - bq_1) - [(-q_2)a + (1 - q_1q_2)b]q_3 \\ &= (1 + q_2q_3)a + (q_1q_2q_3 - q_1 - q_3)b. \end{aligned}$$

Continue this process until r_k is expressed as a linear combination of a and b.

For the numerical example above where 2 = (422, 68) we have (letting $a = 422$ and $b = 68$)

$$\begin{aligned} 14 &= a - b6, \\ 12 &= b - (a - b6)4 = 25b - 4a, \\ 2 &= 14 - 12 = (a - b6) - (25b - 4a) = 5a - 31b. \end{aligned}$$

Thus 2 = (422, 68) = 5(422) − 31(68).

Now that we have established the existence of the g.c.d. of two nonzero integers, we can prove a theorem which states a very important property of primes.

4.3.7 Theorem. *If a prime p divides a product ab, then p divides a or p divides b.*

Proof. Suppose $p \mid ab$ and that $p \nmid a$. Since p is a prime, p and a are relatively prime (or else $a = 1$ and the theorem is proved, since $p \mid ab$). Hence there exist integers r and s such that $1 = pr + as$. Now multiply by b:

$$b = prb + abs.$$

But p divides prb and p divides abs by hypothesis. Thus by 4.2.5, Exercise 2, p divides $prb + abs$. Therefore $p \mid b$.

A simple mathematical induction proves the following corollary.

4.3.8 Corollary. *If a prime p divides a product $a_1 a_2 \cdots a_n$ of integers, then p must divide at least one a_i $(1 \leq i \leq n)$.*

In Definition 4.2.1, we defined "multiple of" as follows: Given two integers a and b, *b is a multiple of a* if there exists an integer c such that $b = ac$. This notion is extended in the following definition.

4.3.9 Definition. *Let a and b be given nonzero integers. The positive integer m is the* ***least common multiple of a and b*** *if the following conditions hold:*

1) m is a multiple of both a and b.
2) If n is a multiple of both a and b, then n is a multiple of m.

Compare this to Definition 4.3.1. We use l.c.m. to abbreviate "least common multiple" and when no confusion can arise, the symbol $[a, b]$ is used to denote the least common multiple of two nonzero integers a and b.

4.3.10 Theorem. *Let a and b be given positive integers. Then the least common multiple of a and b exists; and, in fact,*

$$[a, b](a, b) = ab, \tag{1}$$

where (a, b) denotes the greatest common divisor of a and b.

Proof. We will prove that (1) holds. Since we already know that (a, b) exists, that will prove that $[a, b]$ exists. Let $d = (a, b)$; that is, d is the greatest common divisor of a and b. Then there exist $a_0, b_0 \in I$ such that $a = da_0$ and $b = db_0$. Then by 4.3.11, Exercise 2, $(a_0, b_0) = 1$.

Let f be any common multiple of a and b. Then there exist $r, s \in I$ such that

$$f = ra = ra_0d \qquad \text{and} \qquad f = sb = sb_0d.$$

Thus $b_0d \mid ra_0d$, and hence $b_0 \mid ra_0$. But since $(a_0, b_0) = 1$, 4.3.11, Exercise 6, implies that $b_0 \mid r$; say $r = b_0k$. Then $f = ka_0b_0d$. Now clearly a_0b_0d is a common multiple of a and b and $a_0b_0d \mid f$. Hence the least common multiple of a and b is a_0b_0d. But since $d = (a, b)$, (1) holds as asserted.

Note that this theorem holds if a and b are simply nonzero integers. In this case

$$[a, b](a, b) = \pm ab,$$

since if either a or b is negative, but not both, then ab is negative, whereas $[a, b]$ and (a, b) are positive.

4.3.11 Exercises

1. Find the g.c.d. and l.c.m. of each of the pairs of integers below. Express the g.c.d. as a linear combination of the integers.

   a) 103 and 62 b) 1024 and 361 c) 201 and -1014
   d) 478 and 212

2. If a and b are nonzero integers and $d = (a, b)$ and $a = a_0d$ and $b = b_0d$, prove that $(a_0, b_0) = 1$.

3. If x, y, z, and t are nonzero integers such that $x = yz + t$, show that $(x, z) = (z, t)$.

4. If a and b are nonzero integers and g is a positive integer, then $(ga, gb) = g(a, b)$. Prove it.

5. Prove Corollary 4.3.8.

6. If a and b are nonzero integers such that $(a, b) = 1$ and $a \mid bc$, prove that $a \mid c$.

7. If a and b are nonzero integers such that $d = (a, b)$, then d is precisely the number of integers in the set $\{a, 2a, 3a, \ldots, ba\}$ which are divisible by b. Prove it!

8. Prove that the least common multiple of two positive relatively prime integers is their product.

9. Let a and b be given positive integers and let d be their greatest common divisor. Now on the other hand, consider all common divisors of a and b, that is, all integers which divide both a and b. This set is finite and hence must have a largest element; call it f. Prove that $d = f$. This proves that the greatest common divisor in the sense of our definition is in fact the greatest of the common divisors.

10. Let x, y, and z be nonzero integers.

   a) Prove that $((x, y), z) = (x, (y, z))$.
   b) Define the greatest common divisor of x, y, and z. The symbol for this is (x, y, z).
   c) Prove that $(x, y, z) = ((x, y), z)$.

4.4 THE FUNDAMENTAL THEOREM OF ARITHMETIC

Throughout the beginning courses in algebra and arithmetic, there is one assumption which is made without being stated explicitly. This assumption is that each nonzero integer can be "factored" in exactly one way. In this section we will prove that this assumption is valid; in fact, when it is formalized and proved, it is known as *The Fundamental Theorem of Arithmetic*. The proof will be in two parts.

4.4.1 Theorem. *Every integer $a > 1$ can be expressed as a product of finitely many primes.*

Proof. Here "product" must be interpreted correctly. One prime by itself is to be considered a product. Hence all primes are products of primes. Also 1 is a product of zero primes.

Let S be the set of integers greater than 1 which can *not* be expressed as a product of finitely many primes. The theorem will be proved by showing that S is empty. Assume S is not empty. Then S is a nonempty set of positive integers, and hence must have a smallest element; call it b. Since $b \in S$, b cannot be a prime. Hence there exist integers c and d such that

$$b = cd \quad \text{and} \quad 1 < c < b, \quad 1 < d < b.$$

Therefore, since b is the smallest element of S and $1 < c < b$ and $1 < d < b$, c and d must not be elements of S. That is, c and d can be expressed as a product of finitely many primes; say $c = p_1p_2 \cdots p_n$ and $d = q_1q_2 \cdots q_m$, where $p_1, p_2, \ldots, p_n, q_1, q_2, \ldots, q_m$ are all primes. Then $b = cd = p_1p_2 \cdots p_nq_1q_2 \cdots q_m$, contradicting the fact that $b \in S$. Hence S is empty and the theorem is proved.

If $a > 1$ can be expressed as $a = p_1p_2p_3 \cdots p_n$, where $p_1, p_2, p_3, \ldots, p_n$ are primes, then also $a = p_2p_1p_3 \cdots p_n$ and $a = p_np_1p_2 \cdots p_{n-1}p_3$, etc. But these expressions for a as a product of primes are not essentially different, since exactly the same primes are used in each expression. Consequently, this must be considered in expressing the uniqueness part of the Fundamental Theorem of Arithmetic.

4.4.2 Theorem. *Every integer $a > 1$ can be expressed in only one way as a product of primes except for the order of the factors.*

Proof. As usual, we assume there are two expressions of a as a product of primes; say $a = p_1p_2 \cdots p_n$ and $a = q_1q_2 \cdots q_m$, where $p_1, p_2, \ldots, p_n$, $q_1, q_2, \ldots, q_m$ are primes. We must show that $n = m$ and that the two sets $\{p_1, p_2, \ldots, p_n\}$ and $\{q_1, q_2, \ldots, q_m\}$ are the same. Since $p_1p_2 \cdots p_n = q_1q_2 \cdots q_m$, $p_1 \mid q_1q_2 \cdots q_m$. By Corollary 4.3.8, p_1 divides q_i for some i $(1 \leq i \leq m)$. Hence by 4.2.5, Exercise 7, $p_1 = q_i$. Relabel the q's so that $p_1 = q_1$. Then using the cancellation law of multiplication, we have that $p_2p_3 \cdots p_n = q_2q_3 \cdots q_m$. Thus $p_2 \mid q_2q_3 \cdots q_m$ and so $p_2 = q_j$ for some j $(2 \leq j \leq m)$. Relabel again so that $p_2 = q_2$. By continuing this process one sees that each p_i is equal to a q_i. Hence $\{p_1, p_2, \ldots, p_n\} \subseteq \{q_1, q_2, \ldots, q_m\}$. On the other hand, by reversing the roles of the p's and q's in the above argument, one sees that

$$\{q_1, q_2, \ldots, q_m\} \subseteq \{p_1, p_2, \ldots, p_n\}.$$

Hence these two sets are equal. Therefore $n = m$ and, possibly after a relabeling, $p_i = q_i$ for $i = 1, 2, \ldots, n$.

We now combine these results to obtain the following.

4.4.3 The Fundamental Theorem of Arithmetic. *Every nonzero integer a can be expressed uniquely (except for the order of the factors) in the form*

$$a = (\pm 1)p_1p_2 \cdots p_n,$$

where $p_1, p_2, \ldots, p_n$ are primes.

If $a > 1$, we have already proved the assertion. If $-a > 1$, then by the above, $-a = p_1p_2 \cdots p_n$ for some primes $p_1, p_2, \ldots, p_n$. Hence $a = (-1)p_1p_2 \cdots p_n$. The only other nonzero integers to be considered are 1 and -1. But both 1 and -1 are in the prescribed form where there are no primes in the product.

An alternative method for obtaining uniqueness (rather than ignoring the order of the factors) is to insist on a certain order of the factors. Let a be a nonzero integer. Collect the common terms and express their product by using exponents. Put the "small" primes first. Then a can be expressed as

$$a = (\pm 1)p_1^{n_1}p_2^{n_2} \cdots p_s^{n_s},$$

where $p_1, p_2, \ldots, p_s$ are primes such that $p_1 < p_2 < \cdots < p_s$ and $n_1, n_2, \ldots, n_s$ are nonnegative integers. When a is expressed in this manner, one says that a is expressed in *standard form* as a product of primes.

Let a and b be nonzero integers. By using zero exponents, if necessary, one can express a and b in standard form as products of exactly the *same* primes. Let

$$a = (\pm 1)p_1^{n_1}p_2^{n_2}\cdots p_t^{n_t} \qquad \text{and} \qquad b = (\pm 1)p_1^{m_1}p_2^{m_2}\cdots p_t^{m_t}.$$

Then if for each i $(1 \le i \le t)$, k_i is the minimum of n_i and m_i, and r_i is the maximum of n_i and m_i, then the g.c.d. of a and b is $p_1^{k_1}p_2^{k_2}\cdots p_t^{k_t}$ and the l.c.m. of a and b is $p_1^{r_1}p_2^{r_2}\cdots p_t^{r_t}$. For instance, if $a = 1400$ and $b = 990$, then $a = 2^3\cdot 5^2\cdot 7$ and $b = 2\cdot 3^2\cdot 5\cdot 11$ when they are expressed in standard form. Thus we can write

$$a = 2^3\cdot 3^0\cdot 5^2\cdot 7^1\cdot 11^0 \qquad \text{and} \qquad b = 2^1\cdot 3^2\cdot 5^1\cdot 7^0\cdot 11^1.$$

Hence the g.c.d. of a and b is $2\cdot 5 = 10$, and the l.c.m. of a and b is $2^3\cdot 3^2\cdot 5^2\cdot 7^1\cdot 11^1 = 138{,}600$. This method of finding the g.c.d. of two nonzero integers does not provide the information necessary to write the g.c.d. of two nonzero integers as a linear combination of them.

ADDITIONAL REFERENCES FOR CHAPTER 4

Davenport, H., *The Higher Arithmetic*, London: Hutchinson's University Library (1952).

Hardy, G. H., and E. M. Wright, *An Introduction to the Theory of Numbers*, 4th ed., Oxford: Clarendon Press (1960).

McCoy, Neal H., *Introduction to Modern Algebra*, Boston: Allyn and Bacon (1960).

McCoy, Neal H., *The Theory of Numbers*, New York: Macmillan (1965).

Niven, Ivan, and Herbert S. Zeckerman, *An Introduction to the Theory of Numbers*, 2nd ed., New York: Wiley (1966).

5 groups

The mathematics of today had its beginnings in the early 1800's. New approaches to the concept of number by Carl Friedrich Gauss (German, 1777–1855) and Evariste Galois (French, 1811–1832) and new developments in geometry by N. I. Lobachewsky (Russian, 1793–1856) began the development of modern algebraic and geometric theories in which the *structure of mathematical systems* is the subject of investigations. The creation of mathematical logic by George Boole (English, 1815–1864) in 1847–1854 and the works of David Hilbert (German, 1862–1943), providing geometry with an axiomatic foundation in 1899, were turning points in the passage to postulational methods and the final abstractness of the twentieth century.* Within the last 25 years new advances in mathematics have resulted from more abstraction in the study of categorical algebra.

Much of the mathematics of the twentieth century has become completely formal and abstract. The content of a mathematical theory is the structure of the system determined by the axioms of the system. The language of this theory is provided by the theory of sets, and the rules under which one operates is the mathematical logic. Generalizations and abstractions are obtained by suppressing certain axioms of the number systems and by observing the existence of mathematical systems which satisfy the remaining axioms. Fields, rings, integral domains, and groups are commonly studied abstractions of the number systems.

The addition of real numbers is an associative, commutative binary operation which has an identity element, and each element has an inverse—namely, its negative. Moreover, multiplication of nonzero real numbers satisfies similar conditions. In 2.2.10 we listed some properties

---

* For more information on the history of the development of modern mathematics, refer to E. T. Bell's *Development of Mathematics* (McGraw-Hill, 1945).

of $S(A)$, the set of all one-to-one mappings from a nonempty set A onto itself. Composition of functions is an associative binary operation on $S(A)$ which has an identity element, and every element of $S(A)$ has an inverse in $S(A)$. However, if A has more than two elements, composition of mappings is not commutative on $S(A)$. The recognition of the similarities of these mathematical systems led to the study of groups. By using only the properties common to all such systems and studying the implications of these properties, it is possible to study *all* systems which satisfy these conditions. A system satisfying the properties listed above for function composition on $S(A)$ is known as a *group*. The formal definition of a group given below had been explicitly stated by 1910 and has been the subject of much research in mathematics ever since.

5.1 DEFINITION OF A GROUP

5.1.1 Definition. *A nonempty set of elements G is said to form a* ***group*** *if on G there is defined a binary operation denoted by · satisfying the following axioms.*

G1) *For all $a, b \in G$, $a \cdot b \in G$* *(closure law)*
G2) *For all $a, b, c \in G$, $(a \cdot b) \cdot c = a \cdot (b \cdot c)$* *(associative law)*
G3) *There exists an element $e \in G$ such that $a \cdot e = e \cdot a = a$ for all $a \in G$* *(the existence of an identity element in G)*
G4) *For every $a \in G$, there exists an element $b \in G$ such that $a \cdot b = b \cdot a = e$* *(existence of inverses in G)*

Now let us examine these axioms closely. Axiom G1 is actually redundant. The fact that $\cdot$ is a binary operation on G guarantees that G is closed under $\cdot$. But G1 is included traditionally in the definition of a group to emphasize the closure property of the binary operation. Axiom G2 simply states that the binary operation is associative. In axioms G3 and G4, one must pay particular attention to the quantifiers. Axiom G3 guarantees the existence of an element e, called the identity element, such that $a \cdot e = e \cdot a = a$ for all a in G; that is, e combined with any element a of G is just the element a again. Note that there is one e which must "work" for all a in G. Also, the axiom does not state that e is unique; that is, there may be (as far as we know) more than one element in G which satisfies Axiom G3. Axiom G4 states that, corresponding to each element a in G, there is an element b in G such that $a \cdot b = b \cdot a = e$. The element b may "work" with only this one a. However, each element a of G has such a b. Moreover, axiom G4 does not guarantee the uniqueness of b; thus, for any fixed a in G, there may be several elements b such that $a \cdot b = b \cdot a = e$. The essential difference between the quantification in axioms G3 and G4 is this: G3 is a "there exists, for all" statement, whereas G4 is a "for all, there exists" statement. It is imperative that the reader distinguish between these two ideas. We will prove in the next section that, in fact, the identity element and inverses specified in axioms G3 and G4 are unique.

A group may have many properties other than those specified by the axioms. One property enjoyed by many important groups is that the binary operation is commutative. These groups are so important that we make the following definition.

5.1.2 Definition. *A group G is said to be* ***abelian*** *(or* ***commutative****) if the following axiom is satisfied.*

$$\text{For all } a, b \in G, \ a \cdot b = b \cdot a.$$

The term "abelian group" is most commonly used for groups which also satisfy the axiom above. The name is derived from Niels Henrik Abel (1802–1829), a famous Norwegian mathematician, but its use is so common that "abelian" is usually not capitalized. The reader must be wary of unconsciously assuming that all groups are abelian. Indeed, if A is a set of more than two elements, then $S(A)$ is not abelian.

5.1.3 Examples

We remarked in the introduction that the set of all one-to-one mappings from a nonempty set onto itself forms a group with binary operation as composition of mappings. Let us now see some other familiar examples of groups.

1. Consider the set I of all integers with the usual addition as binary operation. That the closure and associative laws hold is a well-known property of I. Zero is the identity element of I, and for each $n \in I$, $-n$ is its inverse. Hence addition as a binary operation on I satisfies axioms G1 through G4. Hence I is a group. It is also well known that for all $m, n \in I$, $m + n = n + m$. Hence I is an abelian group.

2. Now consider the set I of all integers with the usual multiplication as binary operation. The element 1 is the identity element since, for all $n \in I$, $n \cdot 1 = 1 \cdot n = n$. But there is no element $m \in I$ such that $3 \cdot m = m \cdot 3 = 1$. Hence I with the usual multiplication is *not* a group.

3. This example is quite different from those above. Let G be the set $\{u, v, w, x\}$, and define an operation $\cdot$ on G by the following table.

$\cdot$	u	v	w	x
u	u	v	w	x
v	v	u	x	w
w	w	x	v	u
x	x	w	u	v

The table is used in the following way: To find $v \cdot w$, find v in the vertical column at the left and w in the horizontal row at the top. Then $v \cdot w$ is determined by finding where the row of v crosses the column of w; hence $v \cdot w = x$.

By observation, it is clear that whenever any two elements of G are combined, the result is an element of G. Hence G1 holds. Before considering the associative law, note that u is the identity of G. To verify associativity, one must consider all possible ways of combining three elements of G. However, since u is the identity of G, the investigation may be restricted to all possible ways of combining v, w, and x. But there are still 27 possibilities for which the associativity must be checked. Consequently, the details are not included. All that remains to show that G is a group is to find an inverse for each element of G. However, from the table it is apparent that u and v are their own inverses and w and x are inverses of each other. Hence G is a group. It is left for the reader to show that G is an abelian group.

4. Recall that the set of all rational numbers is the set of all fractions of the form a/b where $a, b \in I$ and $b \neq 0$. Then the set L of all *nonzero* rational numbers forms a group with respect to multiplication. We know that G1 and G2 hold. Since $1 = 1/1$, we have $1 \in L$ and for any $a/b \in L$, $1(a/b) = (a/b)1 = a/b$. Hence G3 holds. Moreover, since for any $a/b \in L$, a/b must be nonzero, then a itself must be a nonzero integer. Thus $b/a \in L$ and $(b/a)(a/b) = (a/b)(b/a) = 1/1 = 1$. Hence G3 holds.

5. Our own experience provides us with the necessary information to prove that the set Q of *all* rational numbers is a group with the usual addition; also that the set R of all real numbers is a group with the usual addition. The proofs that Q and R are groups are left as exercises.

5.2 SIMPLE PROPERTIES OF GROUPS

In axiom G3 of the definition of a group, we asserted the existence of an identity element and remarked that the axiom did not guarantee its uniqueness. In the following theorem, we prove that the identity element of a group is unique. But first, let us examine what this means. That the identity element of a group is unique means that there is exactly one identity element. Since we know by axiom G3 that there is at least one identity element, the negation of "the identity element is unique" is "there exists more than one identity element." Thus, to prove that the identity element is unique, we will assume that there are more than one and then show that they must all be equal. This method of proving uniqueness is, in general, "standard operating procedure."

5.2.1 Theorem. *Let G be a group with binary operation $\cdot$. The identity element of G is unique.*

Proof. Assume that G has two identity elements, e and f. Then, since e is an identity element of G,

(1) $$a \cdot e = e \cdot a = a \qquad \text{for all } a \in G.$$

Similarly, since f is an identity element of G,

(2) $$a \cdot f = f \cdot a = a \qquad \text{for all } a \in G.$$

In particular, since (1) is true for all $a \in G$, it must also be true for $a = f$; that is,

(3) $$f \cdot e = e \cdot f = f.$$

Also (2) must be true for $a = e$; consequently,

(4) $$e \cdot f = f \cdot e = e.$$

By comparing (3) and (4), we see that $e = f$. Hence the identity element of G is unique.

Before proceeding, we will prove the following theorem.

5.2.2 Theorem (Cancellation Laws). *Let a, b, and c be arbitrary elements of a group G with binary operation $\cdot$. Then the following are true.*

i) *If $a \cdot b = a \cdot c$, then $b = c$.*
ii) *If $a \cdot c = b \cdot c$, then $a = b$.*

Proof. We will prove (i) and leave the proof of (ii) as an exercise. Suppose $a \cdot b = a \cdot c$. By axiom G4 of a group, a has an inverse in G;

that is, there exists an element s in G such that $a \cdot s = s \cdot a = e$, where e is the identity element of G. Since $\cdot$ is a binary operation,

$$s \cdot (a \cdot b) = s \cdot (a \cdot c). \tag{5}$$

But

$$\begin{aligned} s \cdot (a \cdot b) &= (s \cdot a) \cdot b && \text{(associative law)} \\ &= e \cdot b && \text{(definition of inverse)} \\ &= b && \text{(definition of identity).} \end{aligned}$$

Similarly,

$$s \cdot (a \cdot c) = (s \cdot a) \cdot c = e \cdot c = c.$$

From (5) and these computations, we see that $b = c$, as asserted, thus proving (i).

Axiom G4 guarantees the existence of an inverse for each element. Let a be an arbitrary element of a group G. Then there exists an element $b \in G$ such that $a \cdot b = b \cdot a = e$, where e is the identity element of G. The next theorem proves that this b is unique. The method of proving uniqueness is the same as that used above.

5.2.3 Theorem. *Let a be an arbitrary element of a group G with binary operation $\cdot$. Then the inverse of a is unique.*

Proof. Assume that b and c are both inverses of a. Then by definition of an inverse of a,

$$a \cdot b = b \cdot a = e \tag{6}$$

and

$$a \cdot c = c \cdot a = e. \tag{7}$$

From (6) and (7) we see that $a \cdot b = a \cdot c$. Hence, by the cancellation law, $b = c$. Thus the inverse of a is unique.

As a consequence of Theorems 5.2.1 and 5.2.3, we can speak of *the* identity element of a group and, for each element a of a group, *the* inverse of a. We will usually use e to denote the identity element of a group, and for the inverse of an element a we usually use the symbol a^{-1}. Hence, for an element a of a group,

$$a \cdot a^{-1} = a^{-1} \cdot a = e.$$

The next two theorems establish some properties of inverses.

5.2.4 Theorem. *If G is a group, then $(a^{-1})^{-1} = a$ for all $a \in G$.*

Proof. Observe that $(a^{-1})^{-1}$ is the inverse of a^{-1}. Hence

$$(a^{-1})^{-1}\cdot(a^{-1}) = (a^{-1})\cdot(a^{-1})^{-1} = e.$$

But since $a\cdot a^{-1} = a^{-1}\cdot a = e$, then a is also the inverse of a^{-1}. Consequently $(a^{-1})^{-1}$ and a are both inverses of a^{-1}. Since the inverse of a^{-1} is unique, we must have $(a^{-1})^{-1} = a$.

5.2.5 Theorem. *If G is a group, then $(a\cdot b)^{-1} = b^{-1}\cdot a^{-1}$ for all $a, b \in G$.*

Proof. By definition $(a\cdot b)^{-1}$ is the inverse of $a\cdot b$. But

$$\begin{aligned}(a\cdot b)\cdot(b^{-1}\cdot a^{-1}) &= \big((a\cdot b)\cdot b^{-1}\big)\cdot a^{-1} && \text{(associative law)}\\ &= \big(a\cdot(b\cdot b^{-1})\big)\cdot a^{-1} && \text{(associative law)}\\ &= (a\cdot e)\cdot a^{-1} && \text{(definition of } b^{-1})\\ &= a\cdot a^{-1} && \text{(definition of } e)\\ &= e && \text{(definition of } a^{-1}).\end{aligned}$$

Similarly,

$$\begin{aligned}(b^{-1}\cdot a^{-1})\cdot(a\cdot b) &= \big((b^{-1}\cdot a^{-1})\cdot a\big)\cdot b = \big(b^{-1}\cdot(a^{-1}\cdot a)\big)\cdot b\\ &= (b^{-1}\cdot e)\cdot b = b^{-1}\cdot b = e.\end{aligned}$$

These computations show that $b^{-1}\cdot a^{-1}$ is the inverse of $a\cdot b$. Hence, since the inverse of $a\cdot b$ is unique, we must have

$$(a\cdot b)^{-1} = b^{-1}\cdot a^{-1}.$$

Since the group G in Theorem 5.2.5 may not be abelian, $b^{-1}\cdot a^{-1}$ may not equal $a^{-1}\cdot b^{-1}$. The reader must be careful not to assume that these are equal.

One method of classifying groups is by consideration of the number of elements in the group. Hence we make the following definition.

5.2.6 Definition. *Let G be a group. If G has a finite number of elements, say n, then we say that the* ***order of G*** *is n. We write this symbolically by $o(G) = n$, and in this case we say G is a* ***finite group.*** *If G does not have a finite number of elements, we say G is an* ***infinite group.***

If a is an element of a group G, then a times itself is $a\cdot a$. If we multiply this by a again, we get $(a\cdot a)\cdot a$. From our past experience with the integers and the real numbers, we expect that there is a shorthand method of denoting the repeated multiplication of a by itself. Hence we make the following definition.

5.2.7 Definition. *Let a be an element of a group G with identity element e. Define*

$$a^0 = e,$$
$$a^1 = a,$$
$$a^2 = a \cdot a,$$

and whenever a^k is defined for a positive integer k, define

$$a^{k+1} = a^k \cdot a.$$

Then, by the Principle of Mathematical Induction, (4.1.2), *a^n is defined for all nonnegative integers n. Furthermore, define*

$$a^{-n} = (a^{-1})^n$$

for all positive integers n. Then a^n is defined for all integers n.

Next we see that this definition of exponents provides the usual rules of exponents.

5.2.8 Theorem. *Let G be a group and let a and b be arbitrary elements of G. Then*

1) $a^m \cdot a^n = a^{m+n}$ for all positive integers m and n,
2) $(a^m)^n = a^{mn}$ for all positive integers m and n, and
3) if G is abelian, $(a \cdot b)^n = a^n \cdot b^n$ for all positive integers n.

Proof. We will prove (1) and leave (2) and (3) as exercises. The proofs of all three statements use mathematical induction. Let S_n be the statement "$a^m \cdot a^n = a^{m+n}$ for all positive integers m." Then S_1 is simply "$a^m \cdot a^1 = a^{m+1}$ for all positive integers m." This is true by the definition of a^{m+1}. Assume that S_k holds for some positive integer k. Then

$$\begin{aligned} a^m \cdot a^{k+1} &= a^m \cdot (a^k \cdot a) && \text{(definition of } a^{k+1}) \\ &= (a^m \cdot a^k) \cdot a && \text{(associative law)} \\ &= a^{m+k} \cdot a && \text{(by } S_k) \\ &= a^{m+k+1} && \text{(definition of } a^{m+k+1}) \end{aligned}$$

for all positive integers m. Hence S_{k+1} is true. Thus, by the Principle of Mathematical Induction, S_n is true for all positive integers n, thus proving (1).

5.2.9 Remark

It is important to observe that the results of Theorem 5.2.8 hold for *all* integers m and n—positive, negative, or zero. Clearly, since $a^0 = e$, the results hold if either m or n is zero. Let us investigate the validity of part (1) in cases in which m or n is a negative integer. Parts (2) and (3) can be handled similarly. Suppose m and n are both negative. Then $-m$ and $-n$ are both positive. Hence,

$$\begin{aligned} a^m \cdot a^n &= (a^{-1})^{-m} \cdot (a^{-1})^{-n} && \text{(definition)} \\ &= (a^{-1})^{-m-n} && \text{(5.2.8 (1))} \\ &= (a^{-1})^{-(m+n)} && \text{(notation)} \\ &= a^{m+n} && \text{(definition).} \end{aligned}$$

Now consider what happens when one exponent is positive and the other is negative. We consider one possibility; the other is similar. Suppose $m > 0$, $n < 0$, and $-n < m$. Let $p = -n$; then p is positive. Hence,

$$\begin{aligned} a^m \cdot a^n &= (a^{m-p} \cdot a^p) \cdot a^{-p} && \text{(5.2.8 (1))} \\ &= a^{m-p} \cdot (a^p \cdot a^{-p}) && \text{(associativity).} \end{aligned}$$

But by a simple induction argument, it can be shown that $a^p \cdot a^{-p} = e$ for all positive integers p. Hence we have $a^m \cdot a^n = a^{m-p} = a^{m+n}$ since $p = -n$.

5.3 EXAMPLES OF GROUPS

In this section we present many examples of groups. A thorough study of each of these examples should develop an intuitive appreciation of groups. In particular, one should note that while all the groups must have the properties required by axioms G1 through G4, whatever additional properties the groups possess may vary widely. The reader should verify that all axioms G1 through G4 hold for each example.

Examples

1. Let G be the set $\{u, v, w, x\}$ and define $\cdot$ by the table

$\cdot$	u	v	w	x
u	x	w	u	v
v	w	x	v	u
w	u	v	w	x
x	v	u	x	w

Is G a group? You may assume that G is associative. What is the identity? What is the inverse of each element? Is it abelian? What is the order of G?

2. Let I be the set of all integers, and let n be a fixed positive integer. We have shown in Section 2.4 that "for $a, b \in I$, $a \equiv b \pmod{n}$ if $a - b$ is a multiple of n" is an equivalence relation on I. For $m \in I$, let $[m]$ denote the equivalence class of m. Let

$$I/(n) = \{[0], [1], [2], \ldots, [n-1]\};$$

i.e., the elements of the set $I/(n)$ are equivalence classes.

Next we show that every element of I belongs to exactly one of the equivalence classes in $I/(n)$. Let k be an integer. By the division algorithm for the integers (Theorem 4.2.3), there exist integers q and r such that $k = qn + r$ and $0 \leq r < n$. Then $k - r = qn$, and so $k \equiv r \pmod{n}$. Thus $k \in [r]$. Since $0 \leq r < n$, then $[r]$ must be one of the elements of $I/(n)$. Hence k belongs to one of the elements of $I/(n)$. Since the elements of $I/(n)$ are distinct equivalence classes, they have no elements in common. This proves our assertion.

Now we define an addition on $I/(n)$. For $[m], [k] \in I/(n)$, set

$$[m] + [k] = [m + k];$$

i.e., the sum of the equivalence class of m and the equivalence class of k is the equivalence class of $k + m$. Hence the sum of the equivalence classes $[m]$ and $[k]$ seems to depend on the integers m and k. But if m_1 and k_1 are integers such that $m_1 \in [m]$ and $k_1 \in [k]$, then $[m_1] = [m]$ and $[k_1] = [k]$. Hence

$$[m] + [k] = [m_1] + [k_1] = [m_1 + k_1].$$

In order to prove that this addition is a binary operation, we must prove that

$$[m_1 + k_1] = [m + k];$$

i.e., for $[m]$ and $[k] \in I/(n)$ there exists a *unique* $[m] + [k] \in I/(n)$. When this has been proved, one says that the addition is *well-defined*. Suppose $[m] = [m_1]$ and $[k] = [k_1]$. Then $m \equiv m_1 \pmod{n}$ and $k \equiv k_1 \pmod{n}$. Hence $m = m_1 + qn$ and $k = k_1 + pn$ for some integers p and q. Then

$$m + k = m_1 + k_1 + (q + p)n,$$

and so

$$m + k \equiv m_1 + k_1 \pmod{n}.$$

Therefore $[m + k] = [m_1 + k_1]$, and so the addition is well-defined.

Finally we prove that $I/(n)$ is a group with the binary operation we have defined. We have already seen that axiom G1 holds. Let

$[m], [k], [q] \in I/(n)$. Then

$$\begin{aligned}([m] + [k]) + [q] &= [m + k] + [q] \\ &= [(m + k) + q] \\ &= [m + (k + q)] \qquad \text{(associative law for } I) \\ &= [m] + [k + q] \\ &= [m] + ([k] + [q]).\end{aligned}$$

Hence G2 holds. Moreover, $[0] \in I/(n)$ and is the identity of $I/(n)$; and for $[m] \in I/(n)$, $[n - m] \in I/(n)$ and $[m] + [n - m] = [n] = [0]$ and $[n - m] + [m] = [n] = [0]$. Hence $[n - m]$ is the inverse of $[m]$. Thus $I/(n)$ is a group. Furthermore, $I/(n)$ is a finite group of order n.

As an illustration, the addition table for $I/(5)$ is given below. For simplification, the brackets [ ] around the elements of $I/(5)$ are omitted; hence 3 means [3], etc.:

+	0	1	2	3	4
0	0	1	2	3	4
1	1	2	3	4	0
2	2	3	4	0	1
3	3	4	0	1	2
4	4	0	1	2	3

3. Let M be the set of all symobols of the form

$$\begin{pmatrix} a & b \\ c & d \end{pmatrix},$$

where $a, b, c, d \in I$. The set M is called the set of 2×2 *matrices over I*. An individual element of M is a *matrix*. Two matrices are *equal* if they are identical; i.e.,

$$\begin{pmatrix} a & b \\ c & d \end{pmatrix} = \begin{pmatrix} e & f \\ g & h \end{pmatrix}$$

if and only if $a = e, b = f, c = g$, and $d = h$. Define an addition on M by setting

$$\begin{pmatrix} a & b \\ c & d \end{pmatrix} + \begin{pmatrix} e & f \\ g & h \end{pmatrix} = \begin{pmatrix} a + e & b + f \\ c + g & d + h \end{pmatrix}$$

for all

$$\begin{pmatrix} a & b \\ c & d \end{pmatrix}, \begin{pmatrix} e & f \\ g & h \end{pmatrix} \in M.$$

Then M is an abelian group with this operation. The identity is

$$\begin{pmatrix} 0 & 0 \\ 0 & 0 \end{pmatrix}$$

and the inverse of

$$\begin{pmatrix} a & b \\ c & d \end{pmatrix} \quad \text{is} \quad \begin{pmatrix} -a & -b \\ -c & -d \end{pmatrix}.$$

Clearly M is an infinite group.

4. Let $x_1, x_2, \ldots, x_n$ be any n symbols. A *permutation* of $x_1, x_2, \ldots, x_n$ is a one-to-one mapping from the set $\{x_1, x_2, \ldots, x_n\}$ onto itself. As was noted before, the set of all permutations of $x_1, x_2, \ldots, x_n$ is a group where the operation is composition of mappings. This group is denoted by S_n and is called the *symmetric group of n symbols.* Even though S_n is a special case of $S(A)$, it is sufficiently important in the study of groups to merit special consideration.

To illustrate the principles involved without the confusion of details, we will consider the case when $n = 4$. Then the mapping α defined by

$$\alpha(x_1) = x_2, \qquad \alpha(x_2) = x_3, \qquad \alpha(x_3) = x_1, \qquad \alpha(x_4) = x_4$$

is a permutation on x_1, x_2, x_3, and x_4. It seems advisable to develop a shorthand notation for the mapping α rather than writing the α and x's so many times. Note that the essential thing about α is that the first symbol is mapped to the second, the second to the third, the third to the first, and the fourth to the fourth. The notation

$$\alpha = \begin{pmatrix} 1 & 2 & 3 & 4 \\ 2 & 3 & 1 & 4 \end{pmatrix}$$

will be used to represent the permutation α. The 1, 2, 3, and 4 of the top row represent the first, second, third, and fourth symbols, respectively, and the bottom row contains their images; i.e., the image of each symbol is directly below it. Hence the permutation

$$\beta(x_1) = x_3, \qquad \beta(x_2) = x_4, \qquad \beta(x_3) = x_1, \qquad \beta(x_4) = x_2$$

could be represented by

$$\beta = \begin{pmatrix} 1 & 2 & 3 & 4 \\ 3 & 4 & 1 & 2 \end{pmatrix}.$$

For ease in discussing these permutations, one usually replaces x_1 by 1, x_2 by 2, etc. Hence α and β are thought of as permutations on the set $\{1, 2, 3, 4\}$.

Since α and β are simply mappings, we can compose them as usual to give $\beta \circ \alpha$; hence

$$\begin{aligned}\beta \circ \alpha &= \begin{pmatrix} 1 & 2 & 3 & 4 \\ 3 & 4 & 1 & 2 \end{pmatrix} \circ \begin{pmatrix} 1 & 2 & 3 & 4 \\ 2 & 3 & 1 & 4 \end{pmatrix} \\ &= \begin{pmatrix} 1 & 2 & 3 & 4 \\ 4 & 1 & 3 & 2 \end{pmatrix}.\end{aligned}$$

To obtain this result, first recall that the mapping on the right must be applied first. What is the image of 1 under this composition? The mapping α takes 1 to 2 and β takes 2 to 4; hence $\beta \circ \alpha$ takes 1 to 4. Similarly, 2 goes to 3, and then 3 goes to 1; hence 2 goes to 1. The element 3 goes to 1, which in turn goes to 3, whereas 4 goes to 4, and then 4 goes to 2. Hence $\beta \circ \alpha$ maps 1, 2, 3, 4 onto 4, 1, 3, 2, respectively. Now what of $\alpha \circ \beta$?

$$\begin{aligned}\alpha \circ \beta &= \begin{pmatrix} 1 & 2 & 3 & 4 \\ 2 & 3 & 1 & 4 \end{pmatrix} \circ \begin{pmatrix} 1 & 2 & 3 & 4 \\ 3 & 4 & 1 & 2 \end{pmatrix} \\ &= \begin{pmatrix} 1 & 2 & 3 & 4 \\ 1 & 4 & 2 & 3 \end{pmatrix}.\end{aligned}$$

This result is obtained as above. It must be kept in mind that the mapping on the right is applied first. Observe that $\beta \circ \alpha \neq \alpha \circ \beta$, since the image of 1 under $\alpha \circ \beta$ is 1, and the image of 1 under $\beta \circ \alpha$ is 4.

Returning to S_n, we can easily see that in general a permutation on the n symbols $x_1, x_2, \ldots, x_n$ can be represented by

$$\begin{pmatrix} 1 & 2 & \cdots & n \\ i_1 & i_2 & \cdots & i_n \end{pmatrix},$$

where $i_1, i_2, \ldots, i_n$ is simply a rearrangement of $1, 2, \ldots, n$. Next let us compute the order of S_n. The order of S_n is the number of ways to map the set $\{1, 2, \ldots, n\}$ onto itself in a one-to-one manner. There are n possible images for 1. After the image of 1 is chosen, there are $n - 1$ possible images for 2, there are $n - 2$ possible images for 3, and so on. Hence the number of permutations of $1, 2, \ldots, n$ is $n(n-1)(n-2)\cdots 3\cdot 2\cdot 1 = n!$ (*n factorial*).

Since S_4 has 4! or 24 elements, it would be extremely tedious to list all the elements and display a table for its binary operation. However, since S_3 has 3! or 6 elements, such an undertaking is feasible. We conclude the discussion of this example by doing just that. The group S_3 is the set $\{\epsilon, \alpha, \beta, \gamma, \delta, \varphi\}$ where

$$\epsilon = \begin{pmatrix} 1 & 2 & 3 \\ 1 & 2 & 3 \end{pmatrix}, \quad \alpha = \begin{pmatrix} 1 & 2 & 3 \\ 2 & 1 & 3 \end{pmatrix}, \quad \beta = \begin{pmatrix} 1 & 2 & 3 \\ 3 & 2 & 1 \end{pmatrix},$$

$$\gamma = \begin{pmatrix} 1 & 2 & 3 \\ 1 & 3 & 2 \end{pmatrix}, \quad \delta = \begin{pmatrix} 1 & 2 & 3 \\ 2 & 3 & 1 \end{pmatrix}, \quad \varphi = \begin{pmatrix} 1 & 2 & 3 \\ 3 & 1 & 2 \end{pmatrix}.$$

Thus,

	ϵ	α	β	γ	δ	φ
ϵ	ϵ	α	β	γ	δ	φ
α	α	ϵ	φ	δ	γ	β
β	β	δ	ϵ	φ	α	γ
γ	γ	φ	δ	ϵ	β	α
δ	δ	β	γ	α	φ	ϵ
φ	φ	γ	α	β	ϵ	δ

5. Consider a rigid square with center 0 and vertices 1, 2, 3, and 4:

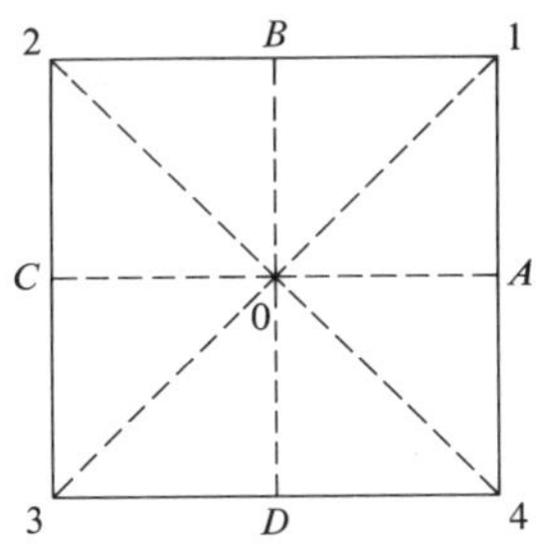

We wish to study all rigid motions of the square into itself; i.e., all motions which return the square to the same position although there may be interchange of some of the vertices. Such a motion can be thought of as a permutation of the symbols 1, 2, 3, and 4. For instance, let ρ denote a 90° counterclockwise rotation about the center. This motion sends 1 to 2, 2 to 3, 3 to 4, and 4 to 1. We can denote this rotation ρ by

$$\rho = \begin{pmatrix} 1 & 2 & 3 & 4 \\ 2 & 3 & 4 & 1 \end{pmatrix}.$$

To perform a 180° counterclockwise rotation, one applies ρ twice, i.e.,

$$\rho^2 = \begin{pmatrix} 1 & 2 & 3 & 4 \\ 3 & 4 & 1 & 2 \end{pmatrix}.$$

Hence,

$$\rho^3 = \begin{pmatrix} 1 & 2 & 3 & 4 \\ 4 & 1 & 2 & 3 \end{pmatrix}$$

and

$$\rho^4 = \begin{pmatrix} 1 & 2 & 3 & 4 \\ 1 & 2 & 3 & 4 \end{pmatrix} = \epsilon,$$

the "identity" motion.

Let σ denote a rotation about the vertical line BD. Then,

$$\sigma = \begin{pmatrix} 1 & 2 & 3 & 4 \\ 2 & 1 & 4 & 3 \end{pmatrix} \quad \text{and} \quad \sigma^2 = \epsilon.$$

Let τ denote a rotation about the horizontal line AC; then

$$\tau = \begin{pmatrix} 1 & 2 & 3 & 4 \\ 4 & 3 & 2 & 1 \end{pmatrix} \quad \text{and} \quad \tau^2 = \epsilon.$$

Let μ and ν denote rotations about the diagonal lines through 1 and 3 and through 2 and 4, respectively. Then

$$\mu = \begin{pmatrix} 1 & 2 & 3 & 4 \\ 1 & 4 & 3 & 2 \end{pmatrix} \quad \text{and} \quad \mu^2 = \epsilon,$$

and

$$\nu = \begin{pmatrix} 1 & 2 & 3 & 4 \\ 3 & 2 & 1 & 4 \end{pmatrix} \quad \text{and} \quad \nu^2 = \epsilon.$$

Now consider the set

$$T = \{\epsilon, \rho, \rho^2, \rho^3, \sigma, \tau, \mu, \nu\}.$$

Prove that T is a group with binary operation being composition of mappings (which we know is associative by Lemma 2.2.8). Make a table for this group.

5.3.1 Exercises

1. Prove Theorem 5.2.2, part (ii).
2. If a, b, and c are elements of a group G, prove that
$$((a \cdot b) \cdot c)^{-1} = (c^{-1}) \cdot b^{-1} \cdot a^{-1}.$$
3. Prove that a group G is abelian if and only if $(ab)^2 = a^2b^2$ for all $a, b \in G$.
4. Prove that a group is abelian if $x^2 = e$ for all x in the group.

5. For each of the following sets, determine whether or not it is a group with the indicated operation. If it is a group, what is the identity element, and what are the inverses of each element?

   a) The set $\{0, 2, 4, 6\}$ of $I/(8)$ with operation addition
   b) The set $\{a + b\sqrt{2} \mid a, b \in I\}$ with operation multiplication
   c) The set $\{a + b\sqrt[3]{2} \mid a, b \in I\}$ with operation multiplication
   d) The set N of nonnegative integers with operation addition
   e) The set I of all integers with operation defined by $a \cdot b = a - b$ for all $a, b \in I$
   f) The set I of all integers with operation defined by $a \cdot b = a + b + 1$
   g) The set I of all integers with operation defined by $a \cdot b = a + b - ab$
   h) The set $\{a, b, c, d\}$ with operation defined by the following table:

$\cdot$	a	b	c	d
a	a	b	c	d
b	b	c	d	b
c	c	d	a	b
d	d	a	c	a

   i) The set $\{a, b, c, d\}$ with operation defined by the following table:

$\cdot$	a	b	c	d
a	d	a	b	c
b	a	b	c	d
c	b	c	d	a
d	c	d	a	b

   Assume that $\cdot$ is associative.

6. Prove parts (2) and (3) of Theorem 5.2.8.

7. Prove that every group which has three, four, or five elements is abelian.

8. Prove that S_3 represents all possible rotations of an equilateral triangle.

9. Consider the group of rotations of the square (Example 5) and the symmetric group S_4. Are they the same group? If not, what is the relationship between these two groups?

In Exercises 8 and 9 above, what would you conjecture about a geometric situation that would produce S_4? Since there must be four vertices, it is natural to guess that all rotations of a tetrahedron would produce S_4. But there is some difficulty in writing it out. Recall that to obtain S_3 by rotating an equilateral triangle, we had to use three-dimensional space. Hence, to visualize the rotations of the three-dimensional tetrahedron, we must use four-dimensional space.

10. Let Q denote the set of rational numbers. Two rational numbers a/b and c/d ($a, b, c, d \in I$, $b \neq 0$, $d \neq 0$) are said to be *equal* if $ad = bc$. Thus $a/b = ax/bx$ for all nonzero $x \in I$. Hence rational numbers have many representations.

    a) The usual addition for rational numbers is

    $$\frac{a}{b} + \frac{c}{d} = \frac{ad + bc}{bd}$$

    for all $a/b, c/d \in Q$. Prove that this addition is well-defined.

    b) Define a new addition on Q by

    $$\frac{a}{b} \oplus \frac{c}{d} = \frac{a + c}{b + d}$$

    for all $a/b, c/d \in Q$. Prove that this addition is not well-defined.

11. Let G_1 and G_2 be two groups. Define an operation on the Cartesian product $G_1 \times G_2$ such that $G_1 \times G_2$ is a group. Include proofs.

12. Let G be a group of order 4. Prove that, except for renaming elements, G has one of the following group tables:

$\cdot$	e	a	b	c
e	e	a	b	c
a	a	b	c	e
b	b	c	e	a
c	c	e	a	b

$\cdot$	e	a	b	c
e	e	a	b	c
a	a	e	c	b
b	b	c	e	a
c	c	b	a	e

5.4 SUBGROUPS

If G is a group, it is possible that there are nonempty subsets of G which are groups themselves with respect to the binary operation defined on G. We say that such subsets are subgroups. More precisely, we make the following definition.

5.4.1 Definition. *A nonempty subset H of a group G is a* **subgroup** *if H is a group with the binary operation of G.*

Note that every group is a subgroup of itself. Moreover, the set consisting of the identity element is a subgroup of every group. Hence every group consisting of more than one element has at least two subgroups. If G is a group with identity element e, then all subgroups of G other than G itself and $\{e\}$ are called *proper* subgroups of G.

To prove that a nonempty subset H of a group G is a subgroup of G, one must verify that H satisfies axioms G1 through G4. However, since the elements of H are also elements of G, they must satisfy axioms G1 through G4 for the group G. We make use of this fact to prove the following theorem, which states that in order to prove that H is a group, it is not necessary to check to see if all four axioms G1 through G4 hold.

5.4.2 Theorem. *Let G be a group with binary operation $\cdot$. A nonempty subset H of G is a subgroup of G if and only if*

1) $a \cdot b \in H$ for all $a, b \in H$, and
2) for each $a \in H$, $a^{-1} \in H$.

Proof. First of all, if H is a subgroup of G, then H is a group itself. Hence (1) and (2) must hold since they are simply G1 and G4 for the group H. Now assume that (1) and (2) hold. We shall prove that H is a subgroup of G by verifying that G1 through G4 hold for H. Since (1) and (2) are true by hypothesis, G1 and G4 hold for H. Let a, b, and c be arbitrary elements in H. Then a, b, and c belong to G, and since G is a group, $a \cdot (b \cdot c) = (a \cdot b) \cdot c$. Hence G2 holds for H. Let g be any element of H. Then (2) guarantees that g^{-1} also belongs to H, and (1) implies that $e = g \cdot g^{-1}$ belongs to H where e is the identity element of H. This proves that G3 holds for H. Therefore H is a group and hence is a subgroup of G.

Next we prove a corollary which makes it even easier to prove that a nonempty subset of a group is a subgroup.

5.4.3 Corollary. *Let G be a group with binary operation $\cdot$. A nonempty subset of H is a subgroup of G if and only if*

3) $a^{-1} \cdot b \in H$ for all $a, b \in H$.

Proof. Suppose (3) holds. To prove that H is a subgroup of G, we will prove that (1) and (2) of Theorem 5.4.2 hold. Let $a \in H$. Then (3) implies that $a^{-1} \cdot a \in H$. Hence $e \in H$. Thus (3) now implies that $a^{-1} \cdot e = a^{-1} \in H$. Thus (2) holds. Now let $a, b \in H$. Then by (2), $a^{-1}, b \in H$. Now (3) implies that $(a^{-1})^{-1} \cdot b \in H$. However, by Theorem 5.2.4, $(a^{-1})^{-1} = a$. Consequently, for $a, b \in H$, we must have $a \cdot b \in H$. We have thus proved that (1) holds, and therefore H is a subgroup of G.

Now, on the other hand, suppose that H is a subgroup of G. Let $a, b \in H$. By G4, $a^{-1} \in H$, and by G1, $a^{-1} \cdot b \in H$. Hence (3) holds.

Either Theorem 5.4.2 or Corollary 5.4.3 may be employed to prove that a certain nonempty subset of a given group is a subgroup of the group. It should be emphasized that all that is required is to verify that (1) and (2) in Theorem 5.4.2 or (3) in Corollary 5.4.3 are satisfied by the elements of the nonempty subset to prove that the nonempty set is a subgroup.

Consider, as an example, the group S_4. The set T of all rotations of a square (Example 5) is a nonempty subset of S_4 and, in addition, T itself is a group. Thus T is a subgroup of S_4. Now consider the subset $U = \{\epsilon, \rho^2, \sigma, \tau\}$ of T. We will prove that U is a subgroup of T. (Note that this will prove that U is also a subgroup of S_4.) To show that U is closed under the binary operation, we exhibit a table:

$\cdot$	ϵ	ρ^2	σ	τ
ϵ	ϵ	ρ^2	σ	τ
ρ^2	ρ^2	ϵ	τ	σ
σ	σ	τ	ϵ	ρ^2
τ	τ	σ	ρ^2	ϵ

Moreover, since each element of U is its own inverse, (1) and (2) of Theorem 5.4.2 are satisfied. Hence U is a subgroup of T.

It should be pointed out that, in general, there is no simple method for finding all subgroups of a given group. One must consider all nonempty subsets, keeping in mind that the subset must be closed under the binary operation of the group and that each element in the subset must have an inverse in the subset. Lagrange's Theorem, which we will prove in Section 5.5, will simplify somewhat the process of finding all subgroups by eliminating some subsets.

5.4.4 Theorem. *Let G be a group with binary operation $\cdot$, and let g be an element of G. Then*

$$H = \{g^i \mid i \text{ is an integer}\}$$

is a subgroup of G.

Proof. Let a and b be arbitrary elements of H. Then $a = g^m$ and $b = g^k$ for some integers m and k. Then $a \cdot b = g^m \cdot g^k = g^{m+k}$ (see Remark 5.2.9). Hence $a \cdot b \in H$. Also $g^{-m} \in H$ and $a \cdot g^{-m} = g^m \cdot g^{-m} = g^{m-m} = e$, by Definition 5.2.7. Similarly $g^{-m} \cdot a = g^{-m} \cdot g^m = e$. Hence g^{-m} is the inverse of a; i.e., $a^{-1} = g^{-m} \in H$. Thus (1) and (2) of Theorem 5.4.2 hold. Therefore H is a subgroup.

There is a special name for groups of this type.

5.4.5 Definition. *A group H is said to be **cyclic** if there exists an element $g \in H$ such that every element of H can be written as g^n for some integer n. In this case, H is called the **cyclic group generated by g**, and g is called **a generator of H.** If H is a subgroup of another group G, then H is called **a cyclic subgroup.***

The subset $V = \{\epsilon, \rho, \rho^2, \rho^3\}$ of the group T of rotations of the square is a subgroup (as is easily checked) and, in fact, is cyclic. The group V can be generated by either ρ or ρ^3. That ρ generates V is evident; to see that ρ^3 also generates V, observe that

$$\begin{aligned}(\rho^3)^1 &= \rho^3,\\ (\rho^3)^2 &= \rho^6 = \rho^4 \cdot \rho^2 = \rho^2,\\ (\rho^3)^3 &= \rho^9 = \rho^8 \cdot \rho = \rho,\end{aligned}$$

and

$$(\rho^3)^4 = \rho^{12} = (\rho^4)^3 = \epsilon.$$

On the other hand, the elements σ, τ, μ, and ν of T generate sub-groups of order 2.

Also, for any positive integer n, the group $I/(n)$ of integers modulo n is cyclic. For example, consider $I/(5)$. Here the binary operation is addition; consequently, if $[x]$ is an element of $I/(5)$, then $[x]^i$ means to *add* $[x]$ to itself i times. Hence, when addition is the binary operation, it is customary to write $i[x]$ instead of $[x]^i$. Though $[1]$ is the most obvious choice of generator for $I/(5)$, it is true in this case that every nonzero element of $I/(5)$ will generate the entire group. Check it! What elements of $I/(6)$ will generate the entire group? What would you conjecture about the generators of $I/(n)$ for an arbitrary positive integer n?

Next we determine the elements in a *finite* cyclic group.

5.4.6 Theorem. *Let G be a finite cyclic group generated by g. If $o(G) = n$, then $g^n = e$ and*

$$\{g, g^2, \ldots, g^{n-1}, g^n = e\}$$

is precisely the set of elements belonging to G.

Proof. First we show that $g^m \neq e$ for any positive integer $m < n$. Suppose this is not the case, that is, suppose $g^m = e$ for some positive integer $m < n$. Let $x \in G$. Then, since G is a cyclic group generated by g, $x = g^k$ for some integer k. Then by the uniqueness part of the division algorithm for the integers (Theorem 4.2.3), there exist unique integers q and r such that $k = qm + r$ and $0 \leq r < m$. Then

$$x = g^k = g^{qm+r} = (g^m)^q \cdot g^r = e^q \cdot g^r = e \cdot g^r = g^r.$$

Hence every element of G is of the form g^r, where $0 \leq r < m$. Hence G has at most m elements, where $m < n$. But by hypothesis, G has exactly n elements, so we have a contradiction, and hence $g^m \neq e$ for any positive integer $m < n$.

We know that the elements $g, g^2, \ldots, g^{n-1}, g^n$ all belong to G. Next we show that these elements are distinct. Suppose that $g^i = g^j$ for some positive integers i and j with $i < j \leq n$. Then, multiplying by g^{-i}, we get

$$e = g^0 = g^{i-i} = g^i \cdot g^{-i} = g^j \cdot g^{-i} = g^{j-i}$$

and since $i < j \leq n$, we must have $0 < j - i < n$. But the first paragraph of this proof showed that $a^m \neq e$ for $m < n$. Hence $g^{j-i} \neq e$, which is a contradiction. Thus the elements $g, g^2, \ldots, g^{n-1}, g^n$ are all distinct.

Since $o(G) = n$, then G must be precisely the set

$$\{g, g^2, \ldots, g^{n-1}, g^n\}.$$

However, since G is a group, e must belong to G. But $g^m \neq e$ if $m < n$. Hence g^n must equal e. This completes the proof of the theorem.

Let us restate the result of this theorem in the following form.

5.4.7 Corollary. *If G is a finite cyclic group generated by g and of order n, then n is the least positive integer such that $g^n = e$.*

Now we are ready to make the following definition.

5.4.8 Definition. *Let g be an element of a group G.*

1) *If there exists a positive integer m such that $g^m = e$, then the smallest such positive integer is the* ***order of g.*** *In this case, g has* ***finite order.***
2) *If no such integer exists, we say g has* ***infinite order.***

We will use the symbol $o(g)$ to denote the order of the element g of G. This should not be confused with the symbol $o(G)$, which denotes the order of the group G.

5.4.9 Remark

Note that Corollary 5.4.7 states that if the cyclic group generated by an element g has order n, then $o(g) = n$. On the other hand, if $o(g) = n$, then, as was shown in the proof of Theorem 5.4.6, $\{g, g^2, \ldots, g^{n-1}, g^n = e\}$ are all distinct. Hence the cyclic group generated by g has order n. Consequently, the order of an element is simply the order of the cyclic group it generates.

The next theorem and corollary tell precisely what the subgroups of a cyclic group are like.

5.4.10 Theorem. *Every subgroup of a cyclic group is cyclic.*

Proof. Let H be a subgroup of a cyclic group G. Let g be the generator of G. If $H = \{e\}$, then H is certainly cyclic with generator e. Hence assume that $H \neq \{e\}$. Since H is a subgroup of G, every element of H is of the form g^k for some integer k. Let m be the least positive integer such that $g^m \in H$. Let g^k be an arbitrary element of H. By the division algorithm for the integers (Theorem 4.2.3), there exist integers q and r such that $k = mq + r$ and $0 \leq r < m$. Then $r = k - mq$, and hence

$$g^r = g^{k+(-mq)} = g^k \cdot g^{-mq} = g^k \cdot (g^m)^{-q}.$$

Since $g^m \in H$, then $(g^m)^{-q}$ must also be in H. Moreover, since $g^k \in H$, then $g^k \cdot (g^m)^{-q}$ must be in H. Hence $g^r \in H$. Now suppose $r \neq 0$; then $0 < r < m$. But since m was chosen to be the least positive integer such that $g^m \in H$, and since $g^r \in H$ with $0 < r < m$, we must have a contradiction. Therefore $r = 0$ and hence $k = mq$. We have shown that an arbitrary element k^k of H is of the form $(g^m)^q$; i.e., H is a cyclic group with generator g^m. This completes the proof.

In proving Theorem 5.4.10, we actually proved the following result, which is more specific about the subgroups of a finite cyclic group.

5.4.11 Corollary. *If a finite cyclic group G has order n and is generated by g, then the subgroups of G are precisely the subgroups generated by g^m where m divides n.*

Proof. By Theorem 5.4.10, we know all of this except that m must divide n. Let H be a subgroup of G. Since $o(G) = n$, we know by Theorem 5.4.6 that $g^n = e$. Hence $g^n \in H$. Now apply the division algorithm, as in the proof of Theorem 5.4.10, using g^n in place of g^k. Thus we see that $n = mq$, and so m divides n, as asserted.

5.4.12 Exercises

1. Prove or disprove that the following sets are subgroups of the corresponding group.
   a) Is the set of all multiples of 6 a subgroup of the group of integers with addition as binary operation?
   b) Is $\{\epsilon, \alpha\}$ a subgroup of the group discussed in Example 4, Section 5.3?
   c) Is $\{3, 6, 9\}$ a subgroup of $I/(10)$ with addition as binary operation?
   d) Is the set of matrices of the form

$$\begin{pmatrix} a & b \\ 0 & 0 \end{pmatrix}$$

   where a and b are integers, a subgroup of the group in Example 3, Section 5.3?

2. Find all subgroups of the following groups.
   a) $I/(8)$ with addition as binary operation
   b) $I/(7)$ with addition as binary operation
   c) The group of rotations of a square; Example 5, Section 5.3 (There are 10 of them.)
   d) S_3, the group of permutations of 3 symbols; Example 4, Section 5.3
3. Let G be a group, and let A and B be subgroups of G. Prove that $A \cap B$ is a subgroup of G. Show that $A \cup B$ need not be a subgroup of G.
4. Let G be an abelian group, and let A and B be subgroups of G. Denote by $A \cdot B$ the set $\{a \cdot b \mid a \in A, b \in B\}$. Prove that $A \cdot B$ is a subgroup of G.
5. Prove that every cyclic group is abelian. By counterexample, show that the converse is not true.
6. Let G be a cyclic group of order 24 and generator g. List *all* subgroups of G.
7. Prove that a subset (nonempty) H of a finite group is a subgroup if and only if H is closed.

5.5 COSETS AND LAGRANGE'S THEOREM

In this section we will study more closely the relationship between a group and its subgroups. In particular, we use a subgroup of a group to perform a decomposition of the group, and we prove the important Theorem of Lagrange, which states that only certain subsets of a group are eligible to be subgroups. The concept of *cosets* is extremely important in the study of modern algebra in general and is not merely a group theoretic notion. This section deserves special attention since the results obtained here will be used throughout the study of algebraic systems.

5.5.1 Definition. *Let G be a group and let H be a subgroup of G. If g is an arbitrary element of G, then*

$$gH = \{g \cdot h \mid h \in H\}$$

is called the ***left coset of H in G determined by g.*** *The element g is often called a* ***representative of the left coset gH.***

Similarly, $Hg = \{h \cdot g \mid h \in H\}$ is called the right coset of H in G determined by g. Of course, if G is abelian, $Hg = gH$ for all $g \in G$. We will consider only left cosets here; however, with minor alterations, the discussion in this section could be repeated for right cosets. Note that if e is the identity element of G, then $eH = H$ so that H itself is a left coset of H in G.

As we will see, the set of all left cosets of a subgroup H in a group G decomposes G into disjoint subsets. Before proceeding with the proof of this, let us consider an example.

In Section 5.3, we defined the group S_3 of permutations on 3 symbols. The elements of S_3 are the following:

$$\epsilon = \begin{pmatrix} 1 & 2 & 3 \\ 1 & 2 & 3 \end{pmatrix}, \quad \alpha = \begin{pmatrix} 1 & 2 & 3 \\ 2 & 1 & 3 \end{pmatrix}, \quad \beta = \begin{pmatrix} 1 & 2 & 3 \\ 3 & 2 & 1 \end{pmatrix},$$

$$\gamma = \begin{pmatrix} 1 & 2 & 3 \\ 1 & 3 & 2 \end{pmatrix}, \quad \delta = \begin{pmatrix} 1 & 2 & 3 \\ 2 & 3 & 1 \end{pmatrix}, \quad \varphi = \begin{pmatrix} 1 & 2 & 3 \\ 3 & 1 & 2 \end{pmatrix}.$$

Let $H = \{\epsilon, \alpha\}$. Then H is a subgroup of S_3. We will compute all the left cosets of H in S_3:

$$\begin{aligned} \epsilon H &= \alpha H = \{\epsilon, \alpha\}, \\ \beta H &= \delta H = \{\beta, \delta\}, \\ \gamma H &= \varphi H = \{\gamma, \varphi\}. \end{aligned}$$

(Check these computations as an exercise.) Hence there are three distinct left cosets of H in S_3. Observe that, in fact, these distinct left cosets form a partition of S_3. We will soon see that this is always the case. Also note that the order of S_3 is the product of the order of H and the number of distinct left cosets of H in S_3.

In Section 2.4 we studied equivalence relations. One important property of equivalence relations that we proved (Theorem 2.4.4) is that the set of equivalence classes forms a partition. Consequently, one should ask if the partition of S_3 by the left cosets of H in S_3 could have been accomplished by an equivalence relation. If so, will this process work in the general case? We answer these questions affirmatively in this section.

5.5.2 Definition. *Let G be a group and H a subgroup of G. For any two elements $a, b \in G$, say* ***a is congruent to b modulo H,*** *and write*

$$a \equiv b \pmod{H},$$

if $b^{-1} \cdot a \in H$.

Thus "congruence modulo H" is a relation on the set G. Here we write $a \equiv b \pmod{H}$ in place of $a \sim b$, which was the symbol for relation in Section 2.4.

5.5.3 Theorem. *The relation defined above is an equivalence relation.*

Proof. We must verify that the reflexive, symmetric, and transitive properties hold. The details are left as an exercise.

5.5.4 Theorem. *Let G be a group with subgroup H. Let $a \in G$ and let $[a]$ denote the equivalence class of a with regard to the equivalence relation $a \equiv b \pmod{H}$ for all $a, b \in G$; that is, $[a] = \{x \in G \mid x \equiv a \pmod{H}\}$. Then*

$$[a] = aH;$$

that is, the equivalence class of a is precisely the left coset of H in G determined by a.

Proof

$$\begin{aligned}[a] &= \{x \in G \mid x \equiv a \pmod{H}\}\\ &= \{x \in G \mid a^{-1}\cdot x = h \text{ for some } h \in H\}\\ &= \{x \in G \mid x = a\cdot h \text{ for some } h \in H\}\\ &= aH.\end{aligned}$$

If one uses right cosets instead of left cosets, one must define $a \equiv b \pmod{H}$ if $a\cdot b^{-1} \in H$ for all $a, b \in G$. Then the equivalence class of $[a]$ is the right coset Ha.

Since the set of left cosets of H in G are simply the equivalence classes of an equivalence relation on the set G, Theorem 1.4.4 immediately gives us the following theorem.

5.5.5 Theorem. *Let G be a group and let H be a subgroup of G. Then the set of left cosets of H in G form a partition of G. Hence every element of G belongs to one and only one left coset of G in H.*

Now we are ready to state and prove the important theorem attributed to the French mathematician Joseph Louis Lagrange (1736–1813).

5.5.6 Theorem (Lagrange). *The order of a subgroup of a finite group divides the order of the group.*

Proof. Let G be a finite group, say $o(G) = n$. Let H be a subgroup of G. Then H is finite, say $o(H) = m$. Let k be the number of left cosets of H in G.

First we show that there is a one-to-one correspondence between any two left cosets of H in G; that is, given two arbitrary left cosets of H in G, say aH and bH, where $a, b \in G$, there exists a one-to-one mapping from aH onto bH. Define a mapping $\alpha: aH \to bH$ by $\alpha(ah) = bh$ for all $h \in H$. Then α is onto since for each $bh \in bH$, there exists $ah \in aH$ such that $\alpha(ah) = bh$. Now let $ah, ah_1 \in aH$, $ah \neq ah_1$. Suppose $\alpha(ah) = \alpha(ah_1)$; that is, $bh = bh_1$. Then, by the cancellation law, $h = h_1$, and so $ah = ah_1$, a contradiction. Hence $\alpha(ah) \neq \alpha(ah_1)$, and therefore α is one-to-one.

Hence each left coset of H in G has the same number of elements. Moreover, since H itself is a left coset, each left coset contains the same number of elements as H; i.e., the number of elements in each left coset is $o(H) = m$. By Theorem 5.5.5, the set of left cosets forms a partition of G. Consequently each element of G lies in exactly one of the k left cosets of H in G. Since each coset has exactly m elements, G must have mk elements. Thus $o(G) = n = mk = o(H)k$. Hence $o(H)$ divides $o(G)$, and the theorem is proved.

Lagrange's Theorem has several corollaries which are important in their own right. We list a few.

5.5.7 Corollary. *If G is a finite group of order n, then $g^n = e$ for all $g \in G$.*

Proof. Let g be an element of G. Then $H = \{g^i \mid i \text{ is an integer}\}$ is a subgroup of G (Theorem 5.4.4); i.e., H is the cyclic subgroup generated by g. By Lagrange's Theorem, $o(H)$ divides $o(G) = n$. Hence, if $o(H) = m$, then $n = mk$ for some positive integer k. By Theorem 5.4.6, $g^m = e$. Hence

$$g^n = g^{mk} = (g^m)^k = e^k = e,$$

which proves the corollary.

5.5.8 Corollary. *The order of every element of a finite group is a divisor of the order of the group.*

Proof. The order of every element of a finite group is the order of the cyclic subgroup it generates, and the order of this subgroup divides the order of the group.

5.5.9 Corollary. *If G is a finite group of order p, where p is a prime integer, then G is cyclic and every element of G except the identity is a generator of G.*

Proof. By Corollary 5.5.8, the order of every element of G except the identity is p, since p has no divisors other than p and 1. Hence the cyclic subgroup generated by any element of G other than the identity has order p and therefore must be all of G.

5.5.10 Definition. *Let G be a finite group and let H be a subgroup of G. The number of left cosets of H in G is the* **index** *of H in G. The symbol $i_G(H)$ denotes the index of H in G.*

By Lagrange's Theorem, if G is a finite group and H is a subgroup of G, then

$$i_G(H) = \frac{o(G)}{o(H)}.$$

5.5.11 Exercises

1. Prove Theorem 5.5.3.
2. Let G be a finite group, and let H_1, H_2 be subgroups of G of orders p, q respectively, where p and q are distinct prime integers. Prove that $H_1 \cap H_2 = \{e\}$.

5.6 NORMAL SUBGROUPS AND FACTOR GROUPS

In this section we consider a special and extremely important collection of subgroups—namely, normal subgroups. The word "normal" here is misleading, as we will see; in fact, normal subgroups are very special and constitute a relatively small class of subgroups. Given a group G and a normal subgroup H of G, we will construct another group related to G and H which will be called the factor group.

5.6.1 Definition. *A subgroup H of a group G is a* ***normal subgroup*** *of G if for all $g \in G$ and for all $h \in H$, $(g \cdot h) \cdot g^{-1} \in H$.*

5.6.2 Theorem. *Let H be a subgroup of a group G. The following are equivalent.*

1) H is a normal subgroup of G.
2) $Ha = aH$ for all $a \in G$; that is, every right coset is also a left coset and vice versa.
3) $aH \cdot bH = (a \cdot b)H$ for all $a, b \in G$; that is, the product of left cosets is also a left coset.

Proof. We will prove that (1) implies (2), (2) implies (3), and (3) implies (1). Note that $Ha = aH$ does not mean that $ha = ah$ for all $h \in H$. The equality $Ha = aH$ simply means that these two sets Ha and aH are the same. Hence, if $Ha = aH$, then for any $h \in H$, there exists $h_1 \in H$ such that $ha = ah_1$. Also $(a \cdot b)H$ denotes the left coset of H in G determined by the element $a \cdot b$ of G.

(1) implies (2). Assume that H is a normal subgroup of G. Let $a \in G$. An arbitrary element of aH is ah where $h \in H$. Since H is a normal subgroup of G, $(a \cdot h) \cdot a^{-1} \in H$; that is, $(a \cdot h) \cdot a^{-1} = h_1$ for some $h_1 \in H$.

Then

$$a \cdot h = (a \cdot h) \cdot (a^{-1} \cdot a) = h_1 \cdot a \in Ha$$

since $h_1 \in H$. Thus $aH \subseteq Ha$. Now pick an arbitrary element $h_2 \cdot a \in Ha$; $h_2 \in H$. Since $a \in G$, $a^{-1} \in G$. Thus, since H is a normal subgroup of G,

$$(a^{-1} \cdot h_2) \cdot (a^{-1})^{-1} = (a^{-1} \cdot h_2) \cdot a \in H.$$

Hence $a^{-1} \cdot (h_2 \cdot a) = h_3$ for some $h_3 \in H$. Then

$$h_2 \cdot a = (a \cdot a^{-1}) \cdot (h_2 \cdot a) = a \cdot h_3 \in aH.$$

Hence $Ha \subseteq aH$, and finally $Ha = aH$.

(2) implies (3). Assume that $aH = Ha$ for all $a \in G$. Let

$$x = (a \cdot h) \cdot (b \cdot h_1) = a \cdot (h \cdot b) \cdot h_1$$

be an arbitrary element of $aH \cdot bH$; that is, $h, h_1 \in H$ and $a, b \in G$. Since $bH = Hb$, there exists $h_2 \in H$ such that $b \cdot h_2 = h \cdot b$. Hence

$$x = a \cdot (b \cdot h_2) \cdot h_1 = (a \cdot b) \cdot (h_2 \cdot h_1) \in (a \cdot b)H$$

since $h_2, h_1 \in H$. Therefore

$$aH \cdot bH \subseteq (a \cdot b)H.$$

Now let $y = (a \cdot b) \cdot h_1$ be an arbitrary element of $(a \cdot b)H$; that is, $a, b \in G$ and $h_1 \in H$. For any $h \in H$,

$$y = (a \cdot b) \cdot h_1 = \big((a \cdot (h \cdot h^{-1}) \cdot b) \cdot h_1\big) = \big((a \cdot h) \cdot (h^{-1} \cdot b)\big) \cdot h_1.$$

But since $bH = Hb$, there exists $h_2 \in H$ such that $h^{-1} \cdot b = b \cdot h_2$. Hence

$$y = (a \cdot h) \cdot \big((b \cdot h_2) \cdot h_1\big) = (a \cdot h) \cdot (b \cdot h_3)$$

for $h_3 = h_2 \cdot h_1 \in H$. Hence $y \in aH \cdot bH$, and thus $(a \cdot b)H \subseteq aH \cdot bH$. Consequently $aH \cdot bH = (a \cdot b)H$.

(3) implies (1). Assume that $aH \cdot bH = (a \cdot b)H$ for $a, b \in G$. Then for any $a \in G$,

$$aH \cdot (a^{-1}H) = (a \cdot a^{-1})H = eH = H.$$

Hence, for $h \in H$, $(a \cdot h) \cdot (a^{-1} \cdot h_1) = h_2$ for some $h_1, h_2 \in H$. Then

$$(a \cdot h) \cdot a^{-1} = \big((a \cdot h) \cdot a^{-1}\big) \cdot (h_1 \cdot h_1^{-1}) = h_2 \cdot h_1^{-1} \in H.$$

Thus H is a normal subgroup of G.

In Section 5.5 we computed the left cosets of the subgroup $H = \{\epsilon, \alpha\}$ in S_3. The distinct left cosets of H in S_3 are H, βH, and γH. Now let us compute the right cosets of H in S_3:

$$H\epsilon = H\alpha = \{\epsilon, \alpha\}, \qquad H\beta = H\varphi = \{\beta, \varphi\}, \qquad H\gamma = H\delta = \{\gamma, \delta\}.$$

By comparing the left and right cosets of H in G, one sees that H is not a normal subgroup of S_3. On the other hand, if $N = \{\epsilon, \delta, \varphi\}$, then N is a subgroup of S_3 and, in fact, N is a normal subgroup of S_3 since the left cosets of N in S_3 are

$$\epsilon N = \delta N = \varphi N = \{\epsilon, \delta, \varphi\} \qquad \text{and} \qquad \alpha N = \beta N = \gamma N = \{\alpha, \beta, \gamma\},$$

whereas the right cosets of N in G are

$$N\epsilon = N\delta = N\varphi = \{\epsilon, \delta, \varphi\} \qquad \text{and} \qquad N\alpha = N\beta = N\gamma = \{a, \beta, \gamma\}.$$

It is property (3) of Theorem 5.6.2 that makes normal subgroups important. It is a tribute to the genius of the French mathematician Evariste Galois* (1811–1832) that he recognized that those groups whose left cosets were also right cosets were special. Using property (3) he was able to define a binary operation on the set of all left cosets of a subgroup in the group. Hence he formed a new group whose elements are cosets.

Let G be a group and let H be a normal subgroup of G. Let G/H be the set of all left cosets of H in G. Define an operation on G/H by

$$aH \cdot bH = (a \cdot b)H \qquad \text{for all } aH, bH \in G/H.$$

We want to prove that this is a binary operation on G/H. By the equivalence of parts (1) and (3) of Theorem 5.6.2, we know that the product of two elements of G/H is also an element of G/H.

However, the multiplication defined in G/H depends on the representatives picked for each left coset; that is, the product of a coset represented by a and a coset represented by b is a coset represented by $a \cdot b$. What if different representatives were picked? For instance, suppose $aH = a_1H$ and $bH = b_1H$. Then

$$aH \cdot bH = (a \cdot b)H \qquad \text{and} \qquad a_1H \cdot b_1H = (a_1 \cdot b_1)H.$$

Is $(a \cdot b)H = (a_1 \cdot b_1)H$? We must prove that $(a \cdot b)H = (a_1 \cdot b_1)H$ in order to prove that the multiplication is a binary operation. In doing so, we will be verifying the "uniqueness" part of the definition of binary operation. One usually says that we are proving that the multiplication is *well-defined.*

---

* Read about Galois' history in E. T. Bell's *Men of Mathematics* (Simon and Schuster, 1937).

Suppose $aH = a_1H$ and $bH = b_1H$ for $a, a_1, b, b_1 \in G$. Then, since these left cosets are equivalence classes for the equivalence relation $a \equiv b \pmod{H}$ if $b^{-1}\cdot a \in H$, we know that

$$a \equiv a_1 \pmod{H} \qquad \text{and} \qquad b \equiv b_1 \pmod{H}.$$

Hence $a_1^{-1}\cdot a \in H$ and $b_1^{-1}\cdot b \in H$; that is, $a_1^{-1}\cdot a = h$ and $b_1^{-1}\cdot b = h_1$ for some $h, h_1 \in H$. Then

$$a = (a_1\cdot a_1^{-1})\cdot a = a_1\cdot h \qquad \text{and} \qquad b = b_1\cdot h_1.$$

This implies that

$$a\cdot b = (a_1\cdot h)\cdot(b_1\cdot h_1) = a_1\cdot\big((h\cdot b_1)\cdot h_1\big).$$

Since H is a normal subgroup of $G, b_1H = Hb_1$. Thus there exists $h_2 \in H$ such that $b_1\cdot h_2 = h\cdot b_1$. Consequently

$$a\cdot b = a_1\cdot\big((b_1\cdot h_2)\cdot h_1\big) = (a_1\cdot b_1)\cdot(h_2\cdot h_1) \in (a_1\cdot b_1)H,$$

since $h_2\cdot h_1 \in H$. Hence $a\cdot b \in [a_1\cdot b_1]$, so that

$$(a\cdot b)H = [a\cdot b] = [a_1, b_1] = (a_1\cdot b_1)H.$$

This proves that the multiplication on G/H is well-defined; i.e., multiplication is a binary operation on G/H.

5.6.3 Theorem. *Let G be a group and let H be a normal subgroup of G. Let G/H be the set of all left cosets of H in G and define*

$$aH\cdot bH = (a\cdot b)H \qquad \text{for all } aH, bH \in G/H.$$

Then G/H is a group with this binary operation.

Proof. We have already verified that the multiplication in G/H is a binary operation on G/H; thus G/H is closed and G1 holds. Now let $aH, bH, cH \in G/H$; that is, $a, b, c \in G$. Then

$$\begin{aligned}(aH\cdot bH)\cdot cH &= (a\cdot b)H\cdot cH\\ &= \big((a\cdot b)\cdot c\big)H\\ &= \big(a\cdot(b\cdot c)\big)H \qquad \text{(since } G \text{ is associative)}\\ &= aH\cdot(b\cdot c)H\\ &= aH\cdot(bH\cdot cH).\end{aligned}$$

Hence the multiplication of G/H is associative, and so G2 holds. $H = eH \in G/H$ and for all $aH \in G/H$,

$$H\cdot aH = (e\cdot a)H = aH$$

and

$$aH\cdot H = (a\cdot e)H = aH.$$

Hence H is the identity of G/H, and so G3 holds. Let $aH \in G/H$. Since G is a group, $a^{-1} \in G$, and so $a^{-1}H \in G/H$. Then

$$aH \cdot a^{-1}H = (a \cdot a^{-1})H = eH = H$$

and

$$a^{-1}H \cdot aH = (a^{-1} \cdot a)H = eH = H.$$

Thus the inverse of aH is $a^{-1}H$. Hence G4 holds, and we have proved that G/H is a group.

5.6.4 Definition. *If G is a group and H is a normal subgroup of G, then the group G/H is called the* ***factor group of G by H.***

5.6.5 Theorem. *Let G be a finite group and let H be a normal subgroup of G. Then G/H is a finite group and*

$$o(G/H) = \frac{o(G)}{o(H)}.$$

Proof. The elements of G/H are precisely the left cosets of G in H. Hence $o(G/H)$ is the index of H in G (Definition 2.5.10). Hence

$$o(G/H) = i_G(H) = \frac{o(G)}{o(H)}.$$

We have shown previously that $N = \{\epsilon, \delta, \varphi\}$ is a normal subgroup of S_3. Then it is possible to form the factor group S_3/N; in fact, $S_3/N = \{N, \alpha N\}$, and its multiplication table is as follows:

$\cdot$	N	αN
N	N	αN
αN	αN	N

Next let us consider an example of a somewhat different type. Let I be the group of integers with addition as binary operation. Let M be the set of all multiples of 4; that is, $M = \{4m \mid m \in I\}$. It is a routine matter to check that M is a subgroup of I. However, since the operation in I is addition, the left cosets of M in I are written as $a + M$ rather than aM. Next observe that the left cosets of M in I are precisely M, $1 + M$, $2 + M$, and $3 + M$; for if $n \in I$, by Theorem 4.2.3, there exist $q, r \in I$ such that $n = 4q + r$ and $0 \leq r < 4$. Hence $n = r + 4q \in r + M$ where $r = 0, 1, 2$, or 3. Furthermore, since I is an abelian group,

$$r + M = \{r + m \mid m \in M\} = \{m + r \mid m \in M\} = M + r.$$

Hence all left cosets of M in I are also right cosets of M in I. Thus M

is a normal subgroup of I. Therefore it is possible to form the factor group I/M.

Hence $I/M = \{M, 1 + M, 2 + M, 3 + M\}$, and the table for the binary operation in I/M follows.

$+$	M	$1 + M$	$2 + M$	$3 + M$
M	M	$1 + M$	$2 + M$	$3 + M$
$1 + M$	$1 + M$	$2 + M$	$3 + M$	M
$2 + M$	$2 + M$	$3 + M$	M	$1 + M$
$3 + M$	$3 + M$	M	$1 + M$	$2 + M$

The two examples of factor groups discussed here have finite order. However, this is not always the case. It is quite possible to have a factor group of infinite order. For instance, the factor group I/K, where I is the group of integers and $K = \{0\}$, has infinite order.

It should be emphasized that the construction of a factor group G/H requires H to be a normal subgroup of G. The factor group G/H plays an extremely important role in the theory of groups. Unfortunately, the use of the factor group in its place in group theory lies beyond the scope of this course. But the distinguishing characteristic of normal subgroups is the fact that being a normal subgroup of a group G allows the construction of the factor group of G.

Before considering mappings on groups, let us consider an example of a group with a geometric flavor. This example is interesting because the abstract algebraic concepts of this chapter can be visualized geometrically.

Let E be the Euclidean plane. Pick a point P in E and put a coordinate system on E with P as origin. Then any point in E can be expressed as an ordered pair (x, y) of real numbers. Define a binary operation on E (considered as a set of points) as follows: Let Q_1 and Q_2 be points in E, say $Q_1 = (x_1, y_1)$ and $Q_2 = (x_2, y_2)$. Then $Q_1 + Q_2 = (x_1 + x_2, y_1 + y_2)$. The following diagram illustrates this addition of points by the construction of the parallelogram with sides PQ_1 and PQ_2:

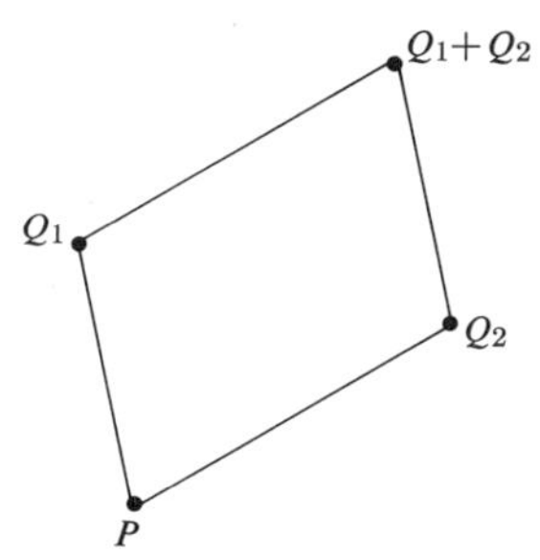

Clearly this operation is closed on E. The associativity follows readily from the associativity of addition of real numbers. The point (0, 0) is the identity element and the inverse of (x, y) is $(-x, -y)$. Thus E is a group with the addition defined above.

What are the subgroups of E? By playing with parallelograms one can readily verify that straight lines through the point P are subgroups of E. This can also be verified analytically. Exercise 8 of 5.6.6 will suggest other choices for subgroups of E. (These exercises are fun and provide a test for your intuitive feeling of groups. Be sure to do them.)

Let H be a straight line in E through P; then H is a subgroup of E. What are the cosets of H in E? One can show that the coset of H in E determined by a point Q is the straight line through Q parallel to H. As Q varies over all points in E, all lines parallel to H are obtained.

Since E is an abelian group, H is a normal subgroup so we can legitimately consider E/H. Consider two elements $Q_1 + H$, $Q_2 + H$ of E/H:

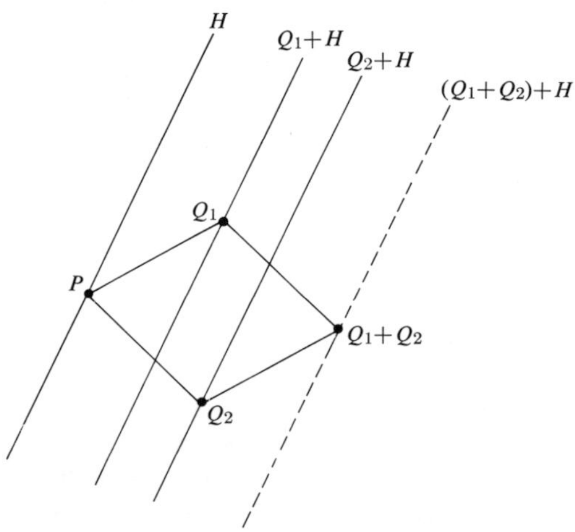

Then $(Q_1 + H) + (Q_2 + H) = (Q_1 + Q_2) + H$. (Recall that these groups are additive.) Since H is a normal subgroup we know *a priori* that this addition in E/H is well-defined. However, for the sake of this example, let us examine the question of well-definedness. Suppose $Q_1 + H = Q_1' + H$ and $Q_2 + H = Q_2' + H$. This means that Q_1 and Q_1' are on the same line parallel to H; similarly Q_2 and Q_2' are on the same line. Analytically one can readily verify that

$$(Q_1 + Q_2) + H = (Q_1' + Q_2') + H.$$

Pictorially the fact that the addition in E/H is well-defined is shown in the following figure:

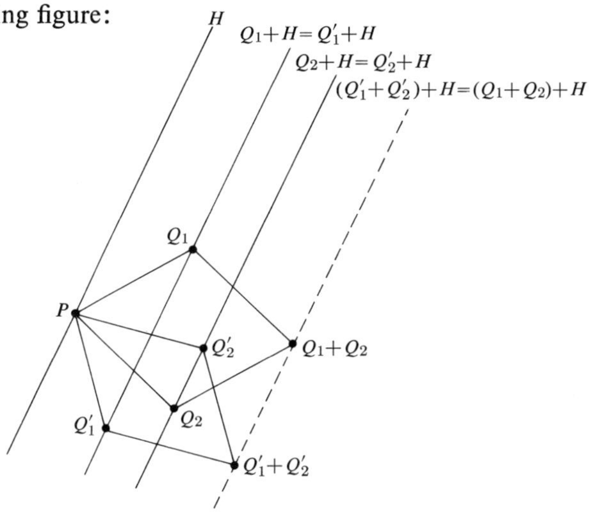

5.6.6 Exercises

Let G be a group.

1. Given that H is a subgroup of G such that $i_G(H) = 2$, prove that H is a normal subgroup of G.
2. Show that the intersection of two normal subgroups of G is a normal subgroup of G.
3. Show that if H is a normal subgroup of G and K is any subgroup of G, then $H \cap K$ is a normal subgroup of K.
4. For any subgroups H and K of G, we define
$$HK = \{h \cdot k \mid h \in H, k \in K\}.$$
   a) Show that if either H or K is a normal subgroup of G, then HK is a subgroup of G.
   b) Show that if both H and K are normal subgroups of G, then HK is a normal subgroup of G.
   c) Show that if H and K are arbitrary subgroups of G, then HK need not be a subgroup of G.
5. Let H be a subgroup of G, and for $g \in G$, define
$$gHg^{-1} = \{(g \cdot h) \cdot g^{-1} \mid h \in H\}.$$
Prove that gHg^{-1} is a subgroup of G.
6. Show that every subgroup of an abelian group is normal.
7. Let H be a subgroup of G, and define $N(H)$ by
$$N(H) = \{g \in G \mid gHg^{-1} = H\}.$$

Then $N(H)$ is called the *normalizer* of H.

a) Prove that $N(H)$ is a subgroup of G.
b) Prove that H is a normal subgroup of $N(H)$.
c) Prove that $N(H)$ is the largest subgroup of G satisfying (b); i.e., prove that if K is a subgroup of G such that H is a normal subgroup of K, then $K \subseteq N(H)$.

8. Let E be the Euclidean plane with the "parallelogram" addition described above.
   a) What is the cyclic subgroup of E generated by a point Q in E different from P?
   b) If the subgroup in (a) is denoted by $\langle Q \rangle$, what are the cosets of $\langle Q \rangle$ in E?
   c) Verify explicitly that the multiplication in $E/\langle Q \rangle$ is well-defined.
   d) Repeat (a) and (b) using a finite number of points $Q_1, Q_2, \ldots, Q_n$ not necessarily on a straight line.
   e) Let S be the set of all points on the x-axis between 1 and 2 inclusive. What is the smallest subgroup of E containing S?
   f) What is the smallest subgroup of E containing a given interval on any line through P?
   g) What is the smallest subgroup of E containing two distinct straight lines through P?
   h) What is the smallest subgroup of E containing a given straight line L through P and a point not on L?

5.7 HOMOMORPHISMS AND ISOMORPHISMS

A group is a nonempty set on which there is defined a binary operation which satisfies axioms G1 through G4. Since a group consists of more than a nonempty set, we should not be surprised that "meaningful" mappings from one group to another should also be more than just mappings. In fact, there should be some interaction between the mapping and the binary operations on the groups. In this section we focus our attention on such mappings from one group to another.

5.7.1 Definition. *Let G and H be groups with binary operations $\cdot$ and $*$ respectively. A* ***homomorphism*** *from G to H is a mapping α from G to H such that*

$$\alpha(g_1 \cdot g_2) = \alpha(g_1) * \alpha(g_2) \tag{1}$$

for all $g_1, g_2 \in G$.

Let us examine condition (1). For elements g_1 and g_2 in G, $g_1 \cdot g_2$ is also in G. Moreover, $\alpha(g_1)$, $\alpha(g_2)$, and $\alpha(g_1 \cdot g_2)$ belong to H. Condition

(1) states that the result is the same whether we combine g_1 and g_2 in G first and then apply the mapping α, or whether we apply the mapping α to g_1 and g_2 separately and then combine $\alpha(g_1)$ and $\alpha(g_2)$ in H. This property of the homomorphism α is often described by saying "α preserves the binary operations of G and H."

A simple example of a homomorphism can be seen by letting G be the cyclic group of order 8 generated by g and letting H be the group of integers modulo 4; i.e.,

$$G = \{e, g, g^2, \ldots, g^7\} \qquad \text{and} \qquad H = \{0, 1, 2, 3\};$$

by letting the binary operation on G be multiplication and on H, addition; and by defining a mapping α from G to H by

$$\begin{array}{llll} \alpha(e) = 0, & \alpha(g) = 1, & \alpha(g^2) = 2, & \alpha(g^3) = 3, \\ \alpha(g^4) = 0, & \alpha(g^5) = 1, & \alpha(g^6) = 2, & \alpha(g^7) = 3. \end{array}$$

That α is a homomorphism can be verified by checking all possible pairs of elements of G to see that condition (1) of the definition of homomorphism holds; i.e., one must verify that $\alpha(g^m \cdot g^k) = \alpha(g^m) + \alpha(g^k)$ for all $g^m, g^k \in G$. To illustrate, one case is verified here:

$$\alpha(g^2 \cdot g^5) = \alpha(g^7) = 3 \qquad \text{and} \qquad \alpha(g^2) + \alpha(g^5) = 2 + 1 = 3.$$

Another example of a homomorphism is provided by the logarithm function (with any fixed base). In 5.1.3, Example 4, we showed that the set L of nonzero rational numbers is a group under multiplication. Also in 5.1.3, Example 5, we mentioned that the set R of all real numbers is a group under addition. The function "$\log_{10} |\ \ |$" is a mapping from L to R such that, for all $x \in L$, $\log_{10} |x|$ means the logarithm to the base ten of the absolute value of x. It is a well-known property of logarithms that for any $x, y \in L$,

$$\log_{10} |xy| = \log_{10} |x| + \log_{10} |y|.$$

Hence $\log_{10} |\ \ |$ is a homomorphism from L to R.

5.7.2 Definition. *Let G and H be groups with binary operations $\cdot$ and $*$ respectively. Let α be a homomorphism from G to H. Then*

1) $Im(\alpha) = \{h \in H \mid \alpha(g) = h$ for some $g \in G\}$ is called the ***image set*** *under α, and*
2) $Ker(\alpha) = \{g \in G \mid a(g) = e\}$, where e is the identity of H, is called the ***kernel*** *of α.*

5.7.3 Theorem. *Let G and H be groups with binary operations $\cdot$ and $*$ respectively. Let e_G and e_H be their identity elements. Let α be a homo-*

morphism from G into H. Then

1) $\alpha(e_G) = e_H$,
2) *for all* $g \in G$, $\alpha(g^{-1}) = (a(g))^{-1}$, *and*
3) $Ker(\alpha)$ *is a subgroup of* G.

Proof
1) Let $g \in G$. Then $g = g \cdot e_G$. Now, applying the homomorphism α, we get

$$\alpha(g) = \alpha(g \cdot e_G) = \alpha(g) * \alpha(e_G).$$

However, since e_H is the identity in H, $\alpha(g) = \alpha(g) * e_H$. Thus

$$\alpha(g) * \alpha(e_H) = \alpha(g) * e_H.$$

Hence, by the cancellation law, $\alpha(e_G) = e_H$.

2) Let $g \in G$. We are to prove that the image of g^{-1} is the inverse of $\alpha(g)$. Since G is a group, $e_G = g^{-1} \cdot g = g \cdot g^{-1}$. Now apply the homomorphism α:

$$e_H = \alpha(e_G) = \alpha(g^{-1} \cdot g) = \alpha(g^{-1}) * \alpha(g)$$

and

$$e_H = \alpha(e_G) = \alpha(g \cdot g^{-1}) = \alpha(g) * \alpha(g^{-1}).$$

Hence $\alpha(g^{-1})$ is the inverse of $\alpha(g)$ in H, and therefore $\alpha(g^{-1}) = (\alpha(g))^{-1}$.

3) Let $g_1, g_2 \in Ker(\alpha)$. Then, by definition of $Ker(\alpha)$, $\alpha(g_1) = e_H$ and $\alpha(g_2) = e_H$. Consequently,

$$\alpha(g_1 \cdot g_2^{-1}) = \alpha(g_1) * \alpha(g_2^{-1}) = e_H * \alpha(g_2)^{-1} = e_H^{-1} = e_H.$$

Hence $g_1 \cdot g_2^{-1} \in Ker(\alpha)$. Thus, by Corollary 5.4.3, $Ker(\alpha)$ is a subgroup of G.

5.7.4 Definition. *Let G and H be groups and let* α *be a homomorphism from G to H.*

1) *If* α *is onto H,* α *is called an* ***epimorphism.***
2) *If* α *is one-to-one,* α *is called a* ***monomorphism.***
3) *If* α *is both an epimorphism and a monomorphism, then* α *is called an* ***isomorphism.*** *If* α *is an isomorphism, then G and H are said to be* ***isomorphic.***

The concept of isomorphism and isomorphic groups is extremely important. When we say two groups G and H are isomorphic, we know that there is a one-to-one mapping from one of the groups onto the other and, in addition, that this mapping preserves the binary operations of the groups. Hence G and H are essentially the same group; their elements may be called different names, but the groups have identically the same structure. Furthermore, if one group has certain additional properties, other than those satisfying G1 through G4, then quite often, because the groups are isomorphic, the second group will also have the additional properties. For instance, (3) of the following theorem states that the property of being abelian is preserved by isomorphisms.

5.7.5 Theorem. *Let G and H be groups and let α be a homomorphism from G to H. Then*

1) *α is an epimorphism if and only if $Im(\alpha) = H$,*
2) *α is a monomorphism if and only if $Ker(\alpha) = \{e_G\}$ where e_G is the identity element of G, and*
3) *if α is an isomorphism, then G is abelian if and only if H is abelian.*

Proof. We will prove (2) and leave (1) and (3) as exercises. Suppose α is a monomorphism and let e_G and e_H be the identity elements of G and H, respectively. By Theorem 5.7.3, $\alpha(e_G) = e_H$. But since α is one-to-one, e_G is the only element of G mapped to the element e_H. Hence $Ker(\alpha) = \{e_G\}$. Conversely, suppose $Ker(\alpha) = \{e_G\}$. Let $g_1, g_2 \in G$ and suppose $\alpha(g_1) = \alpha(g_2)$. Then

$$\begin{aligned} e_H &= \alpha(g_1) * \alpha(g_2)^{-1} \\ &= \alpha(g_1) * \alpha(g_2^{-1}) && \text{(Theorem 5.7.3)} \\ &= \alpha(g_1 \cdot g_2^{-1}) && (\alpha \text{ is a homomorphism}). \end{aligned}$$

Therefore $g_1 \cdot g_2^{-1} \in Ker(\alpha) = \{e_G\}$, and so $g_1 \cdot g_2^{-1} = e_G$. Thus

$$g_1 = g_1(g_2^{-1} \cdot g_2) = (g_1 \cdot g_2^{-1}) \cdot g_2 = e_G \cdot g_2 = g_2.$$

We have shown that $\alpha(g_1) = \alpha(g_2)$ implies $g_1 = g_2$ for all $g_1, g_2 \in G$. Hence α is one-to-one, and so α is a monomorphism.

Homomorphisms play an important role in the development of the theory of any algebraic system. In systems more complex than groups, homomorphisms are still mappings from one set to another which preserve whatever operations are defined on the objects of that system. Homomorphisms in any system provide the means by which objects in that system are compared to one another.

An important example of this role of homomorphisms as a means of comparison is the following theorem, which has extremely wide application in modern abstract algebra.

5.7.6 Theorem (Fundamental Theorem of Homomorphisms). *Let $\alpha: G \to H$ be a homomorphism from a group G onto a group H. Then $G/Ker(\alpha)$ is isomorphic to H.*

Proof. We must define an isomorphism between $G/Ker(\alpha)$ and H. For convenience let $K = Ker(\alpha)$. Recall that elements of G/K are of the form gK for $g \in G$. Define a mapping $\beta: G/K \to H$ by $\beta(gK) = \alpha(g)$ for all $g \in G$.

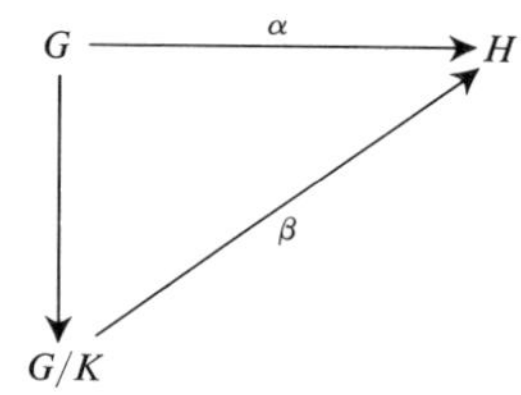

Note that β at least makes sense. We must show that β is well-defined and is, in fact, an isomorphism.

Well-defined. Suppose $g_1K = g_2K$ for $g_1, g_2 \in G$. Since coset formation is an equivalence relation, $g_1g_2^{-1} \in K$. Thus

$$\alpha(g_1)\alpha(g_2)^{-1} = \alpha(g_1)\alpha(g_2^{-1}) = \alpha(g_1g_2^{-1}) = e_G,$$

the identity of G. Thus $\alpha(g_1) = \alpha(g_2)$ and hence $\beta(g_1K) = \beta(g_2K)$.

Homomorphism. Let $g_1K, g_2K \in G/K$. Then

$$\beta(g_1K \cdot g_2K) = \beta(g_1g_2K) = \alpha(g_1g_2) = \alpha(g_1)\alpha(g_2) = \beta(g_1K)\beta(g_2K),$$

so β is a homomorphism.

Epimorphism. Let $h \in H$. Since α is onto, there exists $g \in G$ such that $\alpha(g) = h$. Then

$$gK \in G/K \text{ and } \beta(gK) = \alpha(g) = h.$$

Hence β is onto.

Monomorphism. Suppose $g_1K, g_2K \in G/K$ and $\beta(g_1K) = \beta(g_2K)$. Then $\alpha(g_1) = \alpha(g_2)$. This means that

$$e_G = \alpha(g_1)\alpha(g_2)^{-1} = \alpha(g_1g_2^{-1}).$$

Thus $g_1g_2^{-1} \in K$ and therefore $g_1K = g_2K$. Hence β is one-to-one.

This completes the proof of the theorem.

Exercise 8 below shows an important application of Theorem 5.7.6.

5.7.7 Exercises

1. Define a mapping β from the group M of 2×2 matrices over I to the group I of all integers (i.e., $\beta: M \to I$) by

$$\beta\left(\begin{pmatrix} a & b \\ c & d \end{pmatrix}\right) = a \qquad \text{for all } \begin{pmatrix} a & b \\ c & d \end{pmatrix} \in M.$$

   The binary operation on M and I is addition. Prove that β is a homomorphism. Is β an epimorphism? A monomorphism?

2. Define a mapping δ from I to the group of even integers by $\delta(n) = 2n$ for all $n \in I$. Prove or disprove that δ is a homomorphism.

3. Let α be a homomorphism from a group G to a group H. Prove the following:

   a) $Im(\alpha)$ is a subgroup of H.
   b) $Ker(\alpha)$ is a normal subgroup of G.
   c) $Im(\alpha) = H$ if and only if α is an epimorphism.

4. Let α be an isomorphism from a group G to a group H. Prove that G is abelian if and only if H is abelian.

5. Prove that every cyclic group of order n is isomorphic to $I/(n)$ for every positive integer n.

6. Let H be a normal subgroup of a group G. Define a mapping $\pi: G \to G/H$ by $\pi(g) = gH$ for all $g \in G$. Prove that π is an epimorphism and that $Ker(\pi) = H$. The mapping π is often referred to as the *canonical epimorphism*.

7. Let G and H be subgroups with normal subgroups K and L, respectively. Suppose G/K is isomorphic to H/L. If $o(G) = 24$ and $o(H) = 18$, what is the relationship (if any) between $o(K)$ and $o(L)$?

8. Let H and K be normal subgroups of a group G. Prove that $(H + K)/K$ is isomorphic to $H/(H \cap K)$. [*Hint:* Define a mapping $\alpha: H \to H + K/K$ by $\alpha(h) = hK$ for all $h \in H$. Prove that α is an epimorphism and $Ker(\alpha) = H \cap K$.]

ADDITIONAL REFERENCES FOR CHAPTER 5

Barnes, Wilfred E., *Introduction to Abstract Algebra*, Boston: Heath and Company (1963).

Dixon, John D., *Problems in Group Theory*, New York: Blaisdell (1967).

Herstein, I. N., *Topics in Algebra*, New York: Blaisdell (1964).

McCoy, Neal H., *Introduction to Modern Algebra*, rev. ed., Boston: Allyn and Bacon (1968).

Lang, Serge, *Algebraic Structures*, Reading, Mass.: Addison-Wesley (1967).

Larsen, Max D., *Introduction to Modern Algebraic Concepts*, Reading, Mass.: Addison-Wesley (1969).

Papy, Georges, *Groups*, New York: St. Martin's Press (1964).

Whitesitt, J. Eldon, *Principles of Modern Algebra*, Reading, Mass.: Addison-Wesley (1966).

answers and hints to selected exercises

The enjoyment of mathematics comes from an actual participation in doing mathematics. Mathematics appeals to alert, active minds because there is a challenge which can be mastered through intellectual pursuits. In addition, mathematics is learned through the task of doing mathematics. The concepts are understood only when one has used them in exercises or other applications.

An elaborate section of answers to exercises runs the risk of spoiling the fun for many students and of denying them the opportunity for hard work by which they can learn. Consequently, this section is purposely limited. Some solutions and hints for exercises in Chapters 1 and 2 are given. These exercises were chosen because there may be some bewilderment for beginners over how to proceed. No answers are given for exercises in Chapters 3, 4, or 5. The reason for this is most beautifully expressed by I. N. Herstein in the preface of his book *Topics in Algebra* (Blaisdell, 1964): "Many [exercises] are introduced not so much to be solved but to be tackled. The value of a problem is not so much in coming up with the answer as in the ideas and attempted ideas it forces on the would-be solver." By the time a reader has reached Chapters 3, 4, or 5, he should have matured enough mathematically to be able to tackle the exercises provided there without hints.

1.2.1

1. a) $(q \wedge \sim r) \Rightarrow p$
   b) $(q \Rightarrow p) \wedge \sim r$ [Parts (a) and (b) are confusing, aren't they? They illustrate the weakness of the English language with regard to grouping.]
   c) $(q \Rightarrow \sim r) \wedge (p \Rightarrow r)$

2. a) $\sim(h \Rightarrow r) = h \wedge \sim r$. Here h is "He is happy" and r is "He is rich."
   f) $m \Rightarrow d$. Here m is "He has measles" and d is "He has red dots."

4. a) $\sim p, p \wedge q, p \vee q = \sim(\sim p \wedge \sim q)$,
$p \veebar q = \sim(\sim p \wedge \sim q) \wedge \sim(p \wedge q)$,
$p \Rightarrow q = \sim(p \wedge \sim q), p \Leftrightarrow q = \sim(p \wedge \sim q) \wedge \sim(\sim p \wedge q)$.
b) $\sim p, p \wedge q = \sim(p \Rightarrow \sim q)$, etc.

1.4.29

1. 1.2.1, Exercise 1

a) Symbolically, $\sim[(q \wedge \sim r) \Rightarrow p] = (q \wedge \sim r) \wedge \sim p$; thus the negation is "$q$ and not r and not p" or "q and neither r nor p."

c) $\sim(q \Rightarrow \sim r) \vee \sim(p \Rightarrow r) = (q \wedge r) \vee (p \wedge \sim r)$; thus the negation is "Either q and r or p and not r."

g) $\sim[(p \wedge r) \Rightarrow q] = (p \wedge r) \wedge \sim p$; thus the negation is "$p$ and r and not q."

1.2.1, Exercise 2

a) Being rich is necessary for happiness.

2. Mimic the proof of 1.4.3.

3. Proof of 1.4.8. You fill in the reasons.

$$\begin{aligned} p \Rightarrow q &= \sim p \vee q \\ &= \sim p \vee \sim(\sim q) \\ &= (\sim p \vee \sim p) \vee \sim(\sim q) \\ &= \sim p \vee (\sim p \vee \sim(\sim q)) \\ &= \sim p \vee (\sim(\sim q) \vee \sim p) \\ &= (\sim p \vee \sim(\sim q)) \vee \sim p \\ &= \sim(p \wedge \sim p) \vee \sim p \\ &= p \wedge \sim q \Rightarrow \sim p. \end{aligned}$$

4. Answers are not provided to this exercise because the author does not wish to reduce your feeling of accomplishment. Be sure to do them all.

8. a) $\sim p = p/p$, $p \wedge q = \sim(p/q) = (p/q)/(p/q)$. You do $p \vee q$, $p \veebar q$, $p \Rightarrow q$, and $p \Leftrightarrow q$.

c) If you also took $=$ as undefined, then the (a) part shows that one could develop the same system as we have done in this book. Consequently my only objection would be to the awkwardness and unnaturalness of /.

9. c) Suppose I do not pass this course. Then I must have failed the final exam. But if I had done homework regularly, I would not have failed the final exam. This contradiction establishes the assertion.

1.5.5

1. a) Some men are not mortal.
   b) All men are happy or not rich.
   h) $\sim[(\exists q, q \text{ is eliminated}) \Rightarrow (\exists b, b \text{ is happy})]$
   $$= (\exists q, q \text{ is eliminated}) \wedge \sim(\exists b, b \text{ is happy});$$
   thus the negation is "Some of the girls are eliminated and all the boys are unhappy."

2. a) $\sim[\forall x\, \exists y\; xy = 1] = \exists x\, \forall y\; xy \neq 1$; hence the negation is "There exists x such that $xy \neq 1$ for all y."

1.6.1

These are fun, so they are left for you. Reread the paragraph just before the exercises—it might save you some time.

1.7.1

1. b)

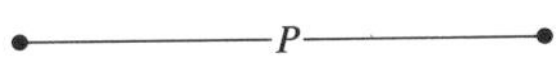

   c)

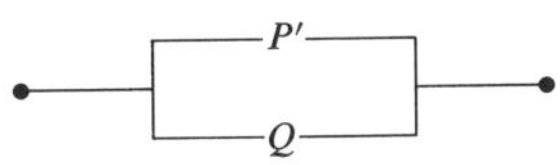

8. *Hint:* Let A_1 be the switch which means "The correct answer to the first question is 'true.'" Then A_1' will mean "The correct answer to the first question is 'false.'" Let A_2 be the switch which means "The student's answer to the first question is 'true.'" Then A_2' is obvious. Let A be the circuit

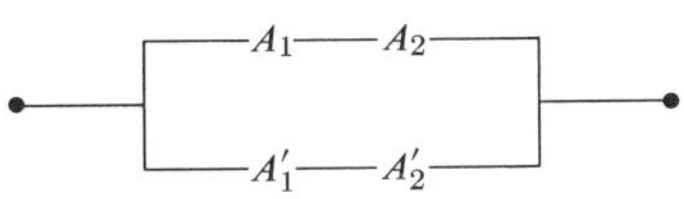

and let A' be the circuit

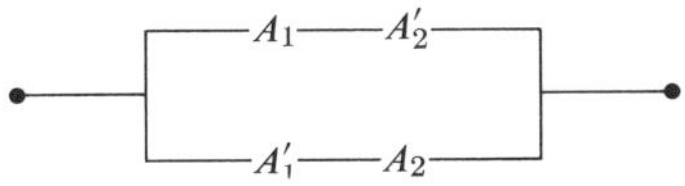

Then A means "The student answers the first question correctly," and A' means "The student answers the first question incorrectly." Construct B, C, D, B', C', and D' similarly for the second, third, and fourth questions. Now design a circuit using these eight small circuits which will operate lights as prescribed in the problem.

2.1.7

6. The first is true; prove it. The second is false; show a counterexample.
7. The solution lies in the following diagram:

Total boys: 24.

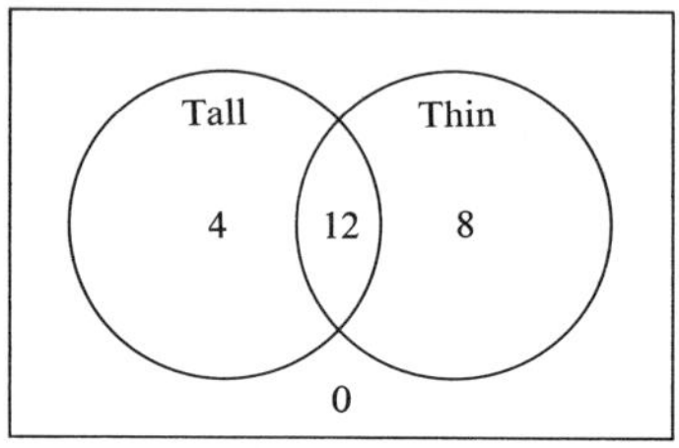

8. Do it like 7.
9. 213; the set A itself, 6 partitions with one set having five elements, 30 partitions with one set having four elements, 100 partitions with one set having three elements, 75 partitions with one set having two elements and no set having more than two elements, and 1 partition of six singletons.

2.2.11

2. b) Neither one-to-one nor onto. The integer 3 is not the image of any integer and 2 and -2 both have the same image.
   d) One-to-one, but not onto. Again 3 is not the image of any integer but $i^3 = j^3$ implies $i = j$, so α is one-to-one.
3. b) Neither
   c) Neither
   d) Onto, but not one-to-one. Recall that every cubic equation with real coefficients has at least one real root; hence ρ is onto. You prove that it is not one-to-one.
4. These are all similar. The pictures should give you ideas for the rest.
   b) Consider the diagram:

P: 1, 2, 3, 4, 5, . . .
↓ ↘ ↘ ↘
E: 2, 4, 6, 8, 10, . . .

Define α by $\alpha(1) = 2$ and $\alpha(n) = 2n + 2$ for $n \geq 2$. [Note that $\beta(n) = 4n$ for all $n \in P$ will also work.]

g)

E: 2, 4, 6, 8, 10, . . .
↓↙ ↙ ↙ ↙
P: 1, 2, 3, 4, 5, . . .

Define α: $E \rightarrow P$ by $\alpha(2) = 1$ and $\alpha(n) = n - \frac{2}{2}$ for all $n \geq 2$. You should check that $n - \frac{2}{2} \in P$ for all $n \in E$, $n \geq 2$.

5. These you must work on. Only hints are given.
   a) What is it you must prove? Suppose α and β are one-to-one. Suppose $(\beta \circ \alpha)(x) = (\beta \circ \alpha)(y)$ for $x, y \in A$. Prove that $x = y$.
   b) Here you must pick an arbitrary element, say c, in C and find an element $a \in A$ such that $(\beta \circ \alpha)(a) = c$. You *must* start in C!

2.3.12

3. If you asked, "What does he mean by 'as many positive integers as negative integers'?", then you have a good question. Interpret it to mean that the two sets are in one-to-one correspondence.

7. Review the argument used in the proof of Theorem 2.3.11.

2.4.5

4. You need to work on these yourself for practice in proving theorems. When proving parts (2) and (4), you must recall that for propositions p and q, $p \Leftrightarrow q = (p \Rightarrow q) \wedge (q \Rightarrow p)$. Therefore you must prove both parts; for instance, in proving (2) you actually prove (i) if $[a] = [b]$ then $a \sim b$ and (ii) if $a \sim b$, then $[a] = [b]$. Also part (4) deals with a negative situation; that is, $[a] \cap [b]$ contains *no* elements and a is *not* equivalent to b. This situation is ideally suited for proof by the indirect method discussed in Section 1.4. Once a part of Theorem 2.4.3 has been proved, it can be used to aid the proof of the remaining parts.

6. Reflexivity is indeed necessary. Doing Exercise 8 will give you a clue in answering Exercise 6.

2.5.8

2. b) This is well-defined. Let

$$\frac{a}{b} = \frac{a_1}{b_1}, \qquad \frac{c}{d} = \frac{c_1}{d_1} \in Q.$$

   Then show that

$$\frac{a}{b} \cdot \frac{c}{d} = \frac{a_1}{b_1} \frac{c_1}{d_1};$$

   this will prove that $\otimes$ is well-defined on Q. You will have to use the definition of equality of rational numbers.

   e) This one is well-defined too. Let $[m] = [m_1]$ and $[n] = [n_1] \in J$. What does this say about m and m_1? About n and n_1? You must prove that $[mn] = [m_1 n_1]$; that is, you must show that $mn - m_1 n_1$ is a multiple of 5.

3. In (a), (b), and (c), you will have to use the associative and commutative properties of the addition and multiplication of the integers.
   d) To show that the usual multiplication of real numbers is associative on T, you must verify that $[(a + b\sqrt{2})(c + d\sqrt{2})](e + f\sqrt{2}) = (a + b\sqrt{2})[(c + d\sqrt{2})(e + f\sqrt{2})]$ for $a, b, c, d, e, f \in I$.

5. These proofs depend on the definition of the intersection of two sets. To prove that $\cdot$ is commutative on $A \cap B$, pick two arbitrary elements say x and y, in $A \cap B$ and prove that $x \cdot y = y \cdot x$. Is it necessary to have $\cdot$ commutative on *both* A and B for $\cdot$ to be commutative on $A \cap B$?

7. Note that (ii) is not an associativity axiom for $+$ on H. You can prove that $+$ is commutative and associative on H and that w is the unique identity element of H.

2.6.6

4. Recall that Q_p is countably infinite. This hint should be all you need.

index

ABCDE79876543210